Six Sigma

and Other
Continuous Improvement Tools
for the Small Shop

Six Sigma
and Other Continuous Improvement Tools
for the Small Shop

Gary Conner

Society of Manufacturing Engineers
Dearborn, Michigan

987654321

Library of Congress Catalog Card Number: 2002108064
International Standard Book Number: 0-87263-583-X

Additional copies may be obtained by contacting:
Society of Manufacturing Engineers
Customer Service
One SME Drive, P.O. Box 930
Dearborn, Michigan 48121
1-800-733-4763
www.sme.org

SME staff who participated in producing this book:
Phillip Mitchell, Senior Editor
Eugene Sprow, Editor
Rosemary Csizmadia, Production Supervisor
Frances Kania, Production Assistant
Kathye Quirk, Graphic Designer/Cover Design
Jon Newberg, Production Editor

Printed in the United States of America

Dedication

To my Creator for allowing me this journey
and to Pamela for sharing her journey with me

Acknowledgments

This book was reviewed for accuracy and relevance to the reader by qualified practioners in the field. Their comments on the text are greatly appreciated. They are individually recognized as follows.

Dennis Miller
Production Team Leader
Neilsen Manufacturing, Inc.
Aumsville, OR

Alan Beeler
Training Coordinator
Neilsen Manufacturing, Inc.
Salem, OR

Mike Begin
Operations Manager
PC Schlosser
Redmond, OR

Terry Begnoche
Senior Product Developer
SME Technical Activities
Dearborn, MI

Daniel Bosserman
Continuous Improvement Coordinator
PED Manufacturing
Oregon City, OR

Scott A. Ford, CMfgT
Black Belt—Six Sigma
Bell Helicopter Textron
Fort Worth, TX

I want to recognize Philip Mitchell and the entire SME staff, including Rosemary Csizmadia, Kathye Quirk, Jon Newberg, and Frances Kania, not only for this effort, but for my first book, *Lean Manufacturing for the Small Shop,* as well.

Special thanks to Phil for helping me focus clearly on the book's objective and developing the concepts into something practical.

I also appreciate the hard work that Eugene put into editing my thoughts into meaningful prose.

Table of Contents

Introduction

"It's a buffet out there!" That's my reaction to the many choices being offered the manufacturing community. Dozens of books each year offer new management theories on how to improve your business. Hundreds of management gurus and consultants stand ready to reorganize how you do everything from purchasing raw materials to turning off the lights at night. Indeed, many of these techniques and technicians could help you achieve major improvements, yet many could also do significant damage if misdirected or misapplied. One common mistake is thinking a small shop can use the exact same tools as the megacorp.

Many proponents are vying for your attention, waving their chosen approach like a red flag, while throwing rocks at others equally enthusiastically waving their own flags. How are you to choose which course is right for you and your company? Does the size of a company really affect the success of a given approach? How does one sort through all this confusion and flag waving?

Unfortunately, a common choice of the small manufacturer is to do nothing, but having gotten you this far, there is hope your response will be much more constructive.

This book is designed to help you sort out the many different approaches, understand them better, and identify the differences, commonality, and synergy these tools can provide for your company. It will not attempt to hold up any one system as the "magic bullet" or end-all solution, but rather show that many of these tools actually overlap and can support each other. In fact, many of these tools can have a powerful synergistic effect, dovetailing together rather than competing for your valuable energies and resources.

SIX SIGMA—ONLY FOR LARGE ORGANIZATIONS?

Six-Sigma consultant Thomas Pyzdek said, "With Six Sigma you are asking for a commitment to an infrastructure. Black belts (the workhorses of a Six-Sigma program) are rotating, full-time positions. We say that black belts should make up about 1% of a company's employment . . . and small businesses would have a hard time doing it." Specifically, he concludes, companies with fewer than 500 employees would therefore struggle with any such implementation (Dusharme 2001).

I strongly disagree. I will admit that dedicating full-time positions to activities such as Lean Manufacturing, Six Sigma, or a World-Class initiative is always a larger challenge for small to medium-sized organizations. However, I look at the positive side of this: the nimble nature of the small shop allows positive changes to happen much more rapidly than in the larger bureaucratic monoliths. If you compare the ratio of hours spent to improvement realized for any major transformation effort, you will find a huge difference in favor of the small company.

So don't allow anyone to discourage you from moving toward a goal of being World Class just because hard work is involved. Being small has major advantages for quick, efficient transformations.

This book will show that the tools of Six Sigma are not in conflict with other World-Class manufacturing tools. Neither are the tools of Six Sigma limited to large firms. Once you understand

and apply these techniques, you and your team can realize incredible performance improvements regardless of your organization's size. A modified or hybrid Six-Sigma approach may be required—based simply on differences in organizational size and complexity—and in each chapter we will try to show examples of how the smaller shop can manage it.

To achieve the level of performance represented by Six Sigma, each and every person in the organization must identify the key metrics for each process. They must then find a way to measure, analyze, improve, and control these metrics. No one can do it for them. Each must play a part in performing this work. Otherwise, it would be like a blindfolded driver getting directions shouted by someone in the back seat.

It is not the responsibility of a Kaizen (continuous improvement) team or outsider black belt or champion to come in and fix your problems. Although everyone wants a quick fix, there is no magic pill that can fix our problems overnight. It takes hard work. It also takes coaching, but remember, the football coach does not play in the game. The coach's job is to help others play the game better. This should be the role of the company black belts: getting everyone working on opportunities for improvement in their own areas. Only then will the company ratchet up in performance quickly and continually, with everyone's working life getting better in the process.

There is always the risk of suboptimization if the interdependency and interaction of different departments is not considered. There should be one shared vision (as in policy deployment to be discussed later) that acts as a common compass, directing all efforts toward the same goal, as opposed to individual optimization at the cost of the whole. We may in fact decide to suboptimize one activity to capitalize on companywide improvements.

LISTEN TO THE VOICE OF THE CUSTOMER

How do you identify the few key metrics that are the most meaningful? One important step is to find out what the critical moment is for your customer. What specific product or service characteristics do they need or want to receive from you? (Note: a customer could be an in-house customer.)

Being too vague about what you need or want to do will always bring vague results. For example, there is a big difference between simply saying, "Our customers want on-time delivery," and saying: "Our customers require delivery no later than 48 hours from the date of order." The latter example is measurable, something that you can analyze, improve, and control.

Applied to the many functions in a job shop or make-to-order shop, each function (that is, purchasing, order entry, engineering, or manufacturing) should be put into words and key indicators established that define the true output requirements from each customer's standpoint. The only way to get these kinds of specifics is to ask your customer what he/she really needs. This is how the process known as "Voice of the Customer" (VOC) was developed, a technique to be discussed later in this book.

By setting up a method to track performance of key dimensions over time, teams develop the ability to define and select the right projects for improvement that will help provide that downstream customer with what they need or want. The team should be coming to the resident black belt or continuous-improvement steering team and asking for help—not to do the project for them, but to assist. Again, not just external customers, but internal customers need to have well defined and measurable quality characteristics provided from their upstream value-chain partners.

For example, an engineering department might phrase their requirement this way: "The fabrication and assembly department requires clear, correct, and complete flat-pattern drawings and assembly schematics at the rate of $120,000 (sales value) per week. Complete drawing packages are due to the production-planning team no later than four working days after receiving customer prints. Questions from the manufacturing or assembly department need to be answered within 1/2 hour. Defects Per Million Opportunities (DPMO) on flat-pattern drawings must be less than 2,500."

What does a DPMO of 2,500 require? If there are 50 dimensions and 200 entities (lines, circles, etc.) on a typical drawing, then there are 250 opportunities per drawing to make a mistake. A DPMO of 2,400 translates to one mistake every 4,000 prints drawn, and if a draftsperson completes an average of 15 drawings per day, then he or she would be allowed only one mistake a year.

A degree in math or statistics is not necessary to be able to measure a process. If you can count defects and divide these into the number of potential opportunities for a defect, you can use this methodology. We will be discussing certain higher-order statistical tools (and pointing toward other books and available resources when you need to use them). But, the real secret to using and benefiting from the Six-Sigma approach is being able to observe a process or the data collected from it and recognize patterns that lead to the identification of an opportunity.

HOW WELL ARE WE MEASURING UP?

Some things obvious to some people are much harder for others to see. Developing stronger observation skills will make it easier to identify opportunities for improvement. Use of Six-Sigma tools can assist in making things visual that were previously hard to recognize.

Software tools are now available for a few hundred dollars that will allow you to enter and analyze data in spreadsheet software packages (you will be directed to locations where you can obtain trial versions). Undoubtedly, you have used many of the basic Six-Sigma tools at some point already. The most important tool in the Six-Sigma toolbox is not a statistical calculator, but the simple question, "Is there a better way of doing this?"

We have been measuring and being measured since early childhood. The report card was one of our first exposures to a measurement tool. Just because something is being measured and recorded, however, is no guarantee that improvement will take place. Behavior often needs to be modified and new procedures implemented before tangible and sustainable improvements can be recognized. A poor report card can generally be traced to a few critical behavior patterns that, once changed, can help restore acceptable results. Although it may be a little late to improve on your report card, this book will deal with tools you can use to find potential causes of desirable and undesirable effects in the processes you now control, in the workplace and elsewhere.

Another critical topic is developing a concise problem statement. A current-condition statement should clearly describe your present situation, along with specific objectives or goals that will help create a framework upon which behaviors (whether human or process) can be measured, tested, adjusted, modified, and controlled.

Whether you are trying to improve a report card or problem solving in a manufacturing process, the steps are not that different. Do not be intimidated by that huge shelf of statistical-process-control books you may have in your library. In this book, the technical nature of these tools is explained where appropriate, but rather than spending time on the theoretical aspect of each statistical function or mathematical formula, common everyday examples are explored where you can apply the tool. Hopefully, this will simplify the application of these tools and make them easier to understand for your entire team.

For the sake of continuity, wherever possible, examples are applied to a small, fictitious company that makes concrete blocks. The position is that Six Sigma works as well in small shops as it does in larger organizations. You don't have to be mass producing circuit boards, washing machines, or motorcycles to benefit from the range of techniques in the Six-Sigma toolbox.

MY CHALLENGE AND COMMITMENT

Writing a book—any book—can be a challenge. You try to match your knowledge, experience, and collection of ideas with the needs of a specific audience. Then you organize it all and try to present it in a style of writing that allows an effective transfer of this information. In this case, I will be trying to reach some who are not exactly avid readers, who at the end of a busy week would simply prefer to relax or just decompress. Sitting down with a technical book is near the bottom of their to-do list. For these folks, I promise to avoid droning on and on about the mundane and theoretical aspect of the tools we will be discussing and focus more on the key aspects of each technique without getting bogged down in micro details.

Much like an automobile owner's manual, we hope that the reader will see this book not only as a tool for strategic planning, but also as a troubleshooting guide and reference for making everyday adjustments and course corrections.

Some readers seeking this information are entrepreneurs, creative individuals who may feel that they already know much of what they need to know about Six Sigma and other continuous improvement tools. The challenge here is to share information about how other entrepreneurs have discovered value in applying these very practical and effectual tools. I promise not to condescend while taking time to explain for the sake of all readers a detail or aspect of a particular technique that may already be familiar to some in this audience.

Some readers may feel they have been hit over the head with every "book-of-the-month" offering on manufacturing improvement since the printing press was invented. If you are in this category, take heart. Don't throw out the tools you have used for years in favor of some new magic bullet. You will be shown how the application of these tools support the efforts you have made over the years. Six Sigma and the rest of the tools described here are no cure-all. Combined with your current skills, effort, knowledge, and experience, I promise to help show you how these tools can be added to those you are already using to greatly enhance your problem-solving ability both professionally and personally.

SIX SIGMA APPLIED TO WRITING A BOOK

Before basic Six-Sigma theory and application are discussed, let me introduce you to some tools from the Six-Sigma toolbox as they could be applied to the writing of this book. After all, I do practice what I preach, so I will demonstrate these tools and the process of continuous improvement by applying them to the writing process.

In Six-Sigma methodology, the basic five-step process used to address a problem or opportunity is called *DMAIC* (pronounced: deh-MAY-ihk). The acronym stands for Define, Measure, Analyze, Improve, and Control.

Define

In writing this book I first need to define the customer and his/her requirements. An effective approach is to take the time to ask potential customers what they need, what they want, and how they want it. I accomplished this by talking to a number of potential customers (readers), some of my clients, and small-business managers. I also spoke with my publisher and editor for their take on customer needs. By accurately defining these needs initially, the scope of this project and the resulting book can be better monitored, measured, and maintained.

Measure

The measurement activity included researching the large number of books and materials available to potential readers of this book. I have over 50 books in my library and the Internet is full of books based on subjects related to our focus here. One key question was "For whom were these current offerings written?" The answer was clearly large manufacturers, not the 10–100-person "Mom-and-Pop" shop. These books are largely theoretical and speak to the chief executive officer (CEO) or vice president (VP) of operations. Using only these "current offerings," a small shop could spend thousands of dollars on books and invest months of time reading to acquire an overall understanding of the many tools available. No single source of information that captures the essence of these many techniques was available in a simple-to-read format.

Analyze

Through brainstorming and independent analysis, an outline was developed to address the problems identified in the measure and define stage. Not every tool known to mankind is identified within this text, however, those that do apply within most job-shop environments have been captured. By categorizing them using the DMAIC formula, we can more meaningfully begin a discussion of the Six-Sigma transformation process.

Improve

In writing this book, I need to ensure that what I generate at the keyboard is *value added*. My objective in writing this book will only be met if my writing provides the reader with an easy-to-interpret tool. It must enable a heightened awareness and greater understanding of the tools while facilitating an easier implementation of the techniques discussed. I want this text to be used as a resource, from which anyone at any level in the organization can gain critical knowledge and examples to more effectively apply these tools to his/her day-to-day responsibilities.

I hope that through self-discovery, clearer understanding, and greater confidence, this book will provide you a larger opportunity to contribute in the process of improvement—that you will see not only the opportunity for improving your company, but the opportunity for self-improvement.

World-Class companies are easy to identify: they are the companies where the majority of the individuals working there are World-Class people. One of the characteristics of World-Class people is that they work hard to develop themselves. They also develop the ability to teach, coach, and lead high-performing teams.

Control

In applying the control idea to the development of this book, a better word might be retention or sustainability. Because people are so busy, any additional information above and beyond what is used in daily life can at times gets lost or be quickly diluted. Thus, it is very important that the information learned be put to use quickly. In fact, the average retention rate of information heard or read for the first time is estimated to be only about 25%. You can improve this through the use of repetition. But, if I chose to use that technique here, this book might end up 800 pages long, and neither of us wants that! Another approach to help ensure retention is to quiz the reader at the end of every section, but there is no guarantee that readers will take the quiz and benefit from it.

In any train-the-trainer program, the best way to build retention is what is called the "teach-back method." For example, assume I just finished teaching you how to set up a machine. After allowing you a chance first to practice, I would then ask you to teach me as if I did not know how to do it. This would not only test how well you learned it, but test my ability at transferring the information or teaching the skill. This process has proven itself to increase the retention rate to about 65% for the first time new skills are learned.

PLEASE TAKE THIS PLEDGE

Another way to cement the relationship here between author and reader is to officially form a contract between us. Please consider the following pledge:

> I, (state your name), promise to share the information gathered from this book with someone else as soon as possible after I read it. This will help me not only retain the information and make my efforts more worthwhile, but begin the process of applying it. I also promise to practice the skills at the first opportunity to further solidify the ideas, insight, and knowledge gained here.

If you have trouble finding someone to talk with about this, I suggest that you send me an e-mail and share your thoughts about what you have learned, or how you hope to implement a technique you read about here. My e-mail address is: Lean1mfg@aol.com. I will also provide you my phone number if you e-mail me a request.

YOUR OWNER'S MANUAL TO SIX SIGMA

Some would argue that a book that covers so many different tools cannot possibly offer the depth of understanding needed to effect the kinds of changes needed in most organizations. You may have heard the phrase "a little knowledge is a dangerous thing!"

As I mentioned earlier, I have some 50 process-improvement books in my personal library, and I add to them every month or so. These books tend to focus on one topic, such as change management, cycle-time reduction, ISO 9000, or Six Sigma. The writers of these books are generally experts on the topics that they write about, having spent years applying a particular tool. Others may have made a life's work of studying and researching these materials. In my case, this book is based on my experience in applying these same tools working with over 50 companies in various industries.

Last winter I finally met one of my life goals. I spent every spare minute rebuilding a truck, pretty much from the ground up. I swapped engines in my 1959 Chevy Apache pickup, put in a new transmission, rebuilt part of the frame, painted it inside and out, installed new bucket seats, and repaired a host of little defects that plagued this 41-year-old vehicle.

Even though I'm relatively satisfied with the job I did, I still consider myself a "shade-tree mechanic," and I typically make more mistakes than I'd like to admit. Part of the problem is that the repair manuals I purchase seldom show information about the exact model I have, and I have to guess a lot or go to the auto-parts store and ask seemingly stupid questions. Having a comprehensive owners' manual showing photographs and examples would really be helpful to shade-tree mechanics like me. This is another purpose for my writing this book: to develop an owner's-manual type of document for continuous improvement. Additionally, I'd like to create a tool that helps to maintain the long-term care of the vehicle (organization) regardless of model (size or industry) and keep it in World-Class shape.

My hope is that this book will serve not as a cover-to-cover kind of book (although reading it through at least once is not a bad idea), but rather as a resource much like a comprehensive owner's manual that we wish would come with every car we buy.

WE CAN'T ALL BE JACK WELCH

Another hurdle I must address is the argument that often comes up about applying World-Class manufacturing techniques to the small shop. You might say to me, "I can see how all this Six Sigma or lean-transformation stuff works at General Electric, Toyota, and a few American auto and motorcycle plants, but we're different. We're just a job shop, we make-to-order. We never know what we will be selling the day after next. We can't get a forecast or plan for next week, much less set up a dedicated cellular-manufacturing system like you describe in your workshop." Is this a valid argument? First, let me say that if you have spoken these words, I feel your pain. You're correct in stating that there is no one-size-fits-all program that will work for every company (or even divisions within a company). You can't put a size 9 shoe on a size 12 foot and expect it to work long term. Similarly you can't pretend to be a Jack Welch running a huge company like GE when your annual sales total is $5 million, not $50 billion. Each organization must approach the goal of becoming World Class in a unique way.

I agree that an exact duplicate of the Motorola or GE approach will not work for the 10-person Mom-and-Pop shop. The black-belt program embraced by larger management teams cannot be adopted verbatim within a 50-person engineer-to-order shop. Yet, many of the techniques do fit perfectly. We will explore as many as possible, and then it is up to you to measure the importance and applicability of each tool to your organization.

Don't look at the list of tools as a buffet or smorgasbord. Granted, some tools are harder to implement than others, and some will not make sense for your company, but do not fall into the trap of avoiding a tool just because it involves hard work. World-Class athletes work hard to improve their performance, and World-Class companies do not shy away from opportunities to improve just because effort is required.

BEGIN WITH THE END IN MIND

In his best seller, *Seven Habits of Highly Effective People,* Stephen Covey performs the difficult task of distilling complex issues down to a single phrase: "Begin with the end in mind." Clearly, for improvement teams to focus their collective efforts on a meaningful objective or vision, we must first verbalize the desired outcome. Visualize it, communicate it, write it down, and articulate it to others.

Setting a goal of perfection also means we should never be satisfied with where we are. We can always improve and set new goals once a milestone has been reached. To "begin with perfection in mind" is to begin with the idea that our company will get as close to perfection as we can bring it, as we continually re-examine the criteria we set for ourselves and establish new goals that bring us even closer.

Every day our competitors are working to surpass us in quality, delivery, service, and cost. If we measure ourselves against them, we will only be motivated to become a little better than they are. If, on the other hand, we set a goal of perfection, our competitors will have the far more difficult (if not impossible) task of trying to catch us. That may be the quintessential definition of Six Sigma: perfection, or something very nearly like it.

Every year or so another World-Class tool or technique is discovered, refined, or repackaged and offered as a cure-all for the manufacturing problems we all face. The truth is, there is no overnight cure for problems that have taken us years to create.

Create? Do we really knowingly create problems? Deliberately create the kinds of headaches we end up dealing with every day? Of course, no one intentionally sets out to design a process or a company that generates waste and unnecessary costs. Yet, just like when we move into a new house, we start out with the best of intentions to keep everything in its proper place. In the garage, power cords, brooms, and yard rakes hang neatly in their respective positions. Check back in a few years, however, and very rarely do those same conditions exist. Slowly, but surely, things migrate away from perfection, and before we know it, we have a big mess on our hands that will take more than a few weekends to restore.

The same dynamic applies to our physical bodies. Obesity is fast becoming the #1 cause of premature death in America. Most of us know this, yet it is hard for us to take the time and put forth the effort to undo years of neglect. To experience real change, *behaviors* need to change. The same is true to apply the tools we will talk about in this book. Changing things may generate some short-term improvement, but only by changing behaviors will we recognize the kind of long-lasting benefits we are seeking. If we narrowly focus on the quick fix, we can expect a quick return, yes, but most likely after that, a fairly quick return of the same problem. If, however, we focus on finding the *root cause* of a problem, then apply the right tool, at the right place, at the right time, the chance for success and sustainability will be substantially increased.

PLAN OF ATTACK

Before beginning, I strongly recommend that you scan the glossary at the back of the book right now for explanations of terms you may have heard before or had some exposure to and need to know more about. A little extra familiarity will help a lot when these terms pop up later. Of course, you can drop back to the glossary at any time, but this will forewarn and forearm you, and I think it is a helpful exercise.

Extremely important to this book is the Troubleshooting Matrix, similar to the one that might be found in a comprehensive automobile owner's manual. I am confident that this matrix will help you identify the right tool at the right time. It is divided into five sections, each dealing with a stage in the DMAIC cycle. If you are looking for help in the define stage, for example, then you can quickly identify tools that will help you with particular problems.

There is a short section dealing with some of the higher-level tools (for example, Theory of Constraints, Baldrige criteria, etc.). It will point out critical overlaps in some of the more acclaimed techniques while building bridges between them and Six Sigma—providing a common language where key tools and techniques complement each other rather than compete for your time and energies.

For some readers there will be tools discussed here that you have used in the past, and so they may be quite familiar. Purposely, every nuance of every tool is not explained, since the goal is to provide an overview and examples for helping you to identify appropriate tools. References and hyperlinks will be provided to assist you in finding additional information.

Chapter 10 describes the need for and the responsibility of each team member to participate. Using the analogy of a light bulb we will talk about the need for every one of us to be generating ideas and shining as brightly as possible. If there are any dark silos (vertically integrated and isolated groups) within an organization, there is little hope of achieving the kind of performance that Six Sigma represents. Also discussed are the dynamics of change, introducing concepts that can make an enormous difference in introducing change and dealing with people who are by nature uncomfortable with it.

REFERENCES

Covey, Stephen R. 1989. *Seven Habits of Highly Effective People.* New York, NY: Simon & Schuster.

Dusharme, Dirk. 2001. "Six-Sigma Survey: Breaking Through the Six-Sigma Hype." *Quality Digest*, November: 27-32.

1

A Zero-sum Game

In *Lean Manufacturing for the Small Shop* (Conner 2001), the principle of the zero-sum game is discussed. For example, five buddies get together to play poker, and each brings $20. After an hour or so, some players have less than $20 and some have more, but the original $100 is still on the table. The total amount in play did not change, it simply changed hands—a zero-sum transaction. This principle applies to business as well.

If one company wins a contract or gains a customer project, it usually means another company lost that contract or project. There is only so much work available, and if a company wants to be on the right side of the zero-sum equation, it needs to be among the best players around. This requires educating management and teams about the nature of the game. It also requires learning about the skills and tools available to the competition. Studying the tools in this book is a good first step. The next step is putting what is learned into practice.

THE INDUSTRIAL REVOLUTION—WHAT WENT WRONG?

Just a few short centuries ago, if you needed a new horseshoe, wagon wheel, or just about anything, it had to be handmade, one-at-a-time. Products in commerce were unique and their components rarely interchangeable. If you needed a replacement part, you had to seek out someone with the skills to make it from scratch.

Then along came Eli Whitney with the idea that mass producing interchangeable parts could vastly improve a product while lowering its cost. Later, Henry Ford applied additional manufacturing refinements and techniques at his Rouge plant. Then came the scientific-management people (Fred Taylor and his followers) with their time-and-motion studies to improve on the tasks each worker performed. Further advances were made by individual manufacturing engineers who designed ever better equipment and processes. This led to the dawning of an entirely new profession, the industrial engineer who focused solely on measuring and improving the work of manufacturing people.

Next came the computer to take some of the more mundane thought processes and calculations away from people and let them spend more time problem solving and decision making. Later came Materials Requirement Planning (MRP), and it became harder for anyone elsewhere in the world to compete with U.S. companies with our mass-production factories overflowing with complex machines, huge inventories, skilled people, and a steady backlog of orders. Manufacturing was king.

Then, in the 1970s and 1980s, the pendulum began swinging the other way. An educated consumer began demanding big improvements in quality, availability, and cost, and some leading U.S. companies began losing business to overseas competitors who were better listeners to these cries of the consumer. Telling customers what they needed to buy (any color as long as it's black) no longer worked. The customer was now calling the shots.

Suddenly, those big-batch superstructures for churning out parts by the millions that had taken a century to build were becoming major liabilities rather than assets. There was a need to become nimble, agile, and flexible again, more like our custom-manufacturing past. Forward-thinking organizations saw themselves at a critical fork in the road, one that threatened extinction if they took the wrong path.

Recognizing the need to change quickly, some companies looked to competitors who were now adopting Just-In-Time, Total Quality Control, and a raft of new methodologies. Others refused to change course and became extinct. (Remember American Motors?) To avoid a similar fate, companies must never rest in their search for improvement and manufacturing perfection.

TAKE A GOOD LOOK IN THE MIRROR

How often when you get up in the morning do you take more than a cursory look in the mirror? You probably don't really want to face up to all those new wrinkles and signs of aging. Businesses also have daily opportunities to look into the figurative mirror, and similarly often fail to pay much attention to what's right there—to see the company as others might see it. What is really being ignored every day is the chance to get better.

One "mirror" that could be looked into is a list of potential waste opportunities like those first identified by Toyota during the early stages of its transformation. Its original list included:

- waste from overproduction,
- waste of waiting time,
- transportation waste,
- processing waste,
- inventory waste,
- wasted motion, and
- waste from defective products.

To this list add the following (and you could probably add even more forms of waste):

- unnecessary paperwork,
- not communicating improvements,
- high nonvalue-added ratio,
- redundant counting,
- inspection after the fact,
- nonfunctional facility layout,
- excessive set-up times,
- unreliable equipment,
- incapable processes,
- lack of maintenance,
- undefined work methods,
- lack of training,
- lack of supervisory ability (coaching),
- lack of workplace organization,
- supplier reliability and quality issues, and
- lack of concern.

How many of the items on this list exist on the floor of the engineering, order-entry, production-planning, and manufacturing-engineering departments in your company? None? If so, put down this book, you are wasting your valuable time. The truth is, hundreds of companies, some with fairly mature continuous-improvement programs, still can realize some major opportunities. This listing can be used as a mirror—paying particular attention to where major adjustments and improvements need to be made—to begin the journey to transform your company from "wasteful" to "World Class."

HOW IMPORTANT IS QUALITY?

Some people mistakenly assume that Six Sigma is only associated with statistical process control, the program credited with positive change at Motorola and General Electric. These efforts are generally viewed as quality initiatives, meant to control the quality of product as it flowed through their manufacturing processes. Although process-control quality is certainly an element of the Six-Sigma approach, such a narrow view of its principles and concepts understates the potential for applying it across all company activities.

How important has quality been throughout history? Egyptian engineers and builders were required to have their measurement device (the cubit stick) calibrated every full moon, and if they failed to do so, the penalty could include being put to death. (Maybe that's why the pyramids are still standing.) Roman engineers also had the threat of death hanging over their head if a building they designed or constructed fell down and killed someone. Many people are unaware that one factor contributing to the French Revolution was the people in power abusing quality standards—buying gold with one set of scales and selling it with a different scale that was more profitable for them. Today, watchdog groups monitor businesses and governmental agencies to avoid such blatant abuses. Quality and fairness are clearly important. Though companies may not face going out of business for manufacturing mistakes, major lawsuits can be filed for negligence if a defect gets passed on to a customer and causes injury.

How reasonable or possible is it to assure total perfection in what is produced? Can a process be created that has no potential to manufacture a defective product? What is an acceptable limit of defects? How about 99%? Not good enough? Would you be satisfied with a process that was 99.9% reliable—only one defect per 1,000 opportunities?

Most people feel 99.9% is pretty good odds against failure. Yet, here are some bad examples from a variety of industries of what would happen if they were allowed no more than one defect per 1,000 opportunities:

- 2 million documents lost by the IRS each year;
- 22,000 checks deducted from the wrong checking account in the next hour;
- 12 babies given to the wrong parents every day;
- 2,488,200 books printed with the wrong cover each year;
- 291 pacemaker implants performed incorrectly each year; and
- 114,500 mismatched pairs of shoes shipped each year.

Do most people get it right 99.9% of the time? It makes you wonder doesn't it? Is the answer to add inspectors to ensure that 100% of everything manufactured is checked? The truth is, 100 inspectors could be added, all looking at the same product 100% of the time, and still defects would be missed.

Here is a little exercise. Take one minute to read the paragraph in Figure 1-1. Count the number of times the letter "F" appears in the text. In this exercise, "F" represents a defect.

How many Fs did you count? There are 17 Fs in the paragraph, and most people find only 12 of them. This is partly because we can't help getting caught up in the story, and partly because we "see" what we hear. We read the word "of" and say "uv" to our inner ear and do not hear or see the F.

> From the Bainbridge ferry, they look like raspberry smears against the grunge gray of the south Seattle waterfront. Some may wonder why anyone would use the same color of paint as Marilyn Monroe's lipstick to paint ships. But there they are, hoisted up in dry dock showing the ugly, bulging bottoms that led a former officer to call them pregnant guppies. Of all the red fleet tied up at Pier 36, 100% are earmarked to go out and be purposely run into large pieces of ice. Seattle is a city of superlatives—during the Yukon Gold Rush, it was known as the only city ever to own a territory, and the Smith Tower was one of the tallest buildings west of the Mississippi. The bright-red ships add one more of these titles. By default and precedent, Seattle has become the nation's polar ice-breaking capital. It's the only place where the Coast Guard bases two of its icebreakers going to the Arctic and Antarctic.

Figure 1-1. Count the Fs exercise.

If you were assigned to read a book with 1 million Fs, you would have 1 million opportunities to miss one. A reader able to catch an average of 999,996.6 (missing only 3.4) would be operating at Six-Sigma levels. How did you do on the reading assignment? The same paragraph is shown in Figure 1-2, with Fs replaced with Xs to show them more clearly.

Xrom the Bainbridge Xerry, they look like raspberry smears against the grunge gray oX the south Seattle waterXront. Some may wonder why anyone would use the same color oX paint as Marilyn Monroe's lipstick to paint ships. But there they are, hoisted up in dry dock showing the ugly, bulging bottoms that led a Xormer oXXicer to call them pregnant guppies. OX all the red Xleet tied up at Pier 36, 100% are earmarked to go out and be purposely run into large pieces oX ice. Seattle is a city oX superlatives—during the Yukon Gold Rush, it was known as the only city ever to own a territory, and the Smith Tower was one oX the tallest buildings west oX the Mississippi. The bright-red ships add one more oX these titles. By deXault and precedent, Seattle has become the nation's polar ice-breaking capital. It's the only place where the Coast Guard bases two oX its icebreakers going to the Arctic and Antarctic.

Figure 1-2. Counting exercise with Xs instead of Fs.

Based on the number of Fs you missed, Table 1-1 shows how many Fs (defects) you would overlook if your assignment was to read a book with a million Fs (defect opportunities). Say, for example, you counted 14 Fs, missing three. Looking up three missed in the matrix, you can see that if faced with a million opportunities to miss an F, there is a good chance you would miss 176,471 of them.

Don't feel bad if you did not get all 17. Everyone is human, and we all make mistakes. The goal in trying to attain Six-Sigma performance levels must include developing processes that make it very hard to make a mistake. The discussion of Poka-Yoke (mistake proofing) will deal more with the issue of process design. There will always be process-control opportunities for enhancing the consistency of your company's product quality.

Although there is no cost to missing an F in the exercise, what is the cost of a defect generated in your manufacturing or service

Table 1-1. DPMO chart for F-counting exercise

Missed	Defects Per Million (DPM)
0	0
1	58,824
2	117,647
3	176,471
4	235,294
5	294,118
6	352,941
7	411,765
8	470,588
9	529,412
10	588,235
11	647,059
12	705,882
13	764,706
14	823,529
15	882,353
16	941,176
17	1,000,000

business? Chances are, it will depend on when the defect is caught, and whether you catch it or your customer catches it.

Cost of Poor Quality (COPQ) is a term companies on the path toward World-Class performance should recognize. Even though there may be differences in people's experience with statistics or business-management techniques, one commonality in every group is an understanding of money and its value. Making the Defects Per Million Opportunities (DPMO) stand out by putting a dollar figure to the cost of an average defect and then multiplying that by the current DPMO can produce an astonishing figure that quickly gets people's attention.

TROUBLESHOOTING TOOLS

Many tools will be introduced in this text, and they might get a little confusing without some forewarning and explanation. Table 1-2 is a troubleshooting guide that shows how the tools interrelate and which troubles they are best at solving. Take a moment to study this because it is the key to your master plan and illustrates how these tools fit within the Six-Sigma umbrella.

In an effort to maintain a common approach, each tool described will follow the model shown here (along with any additional sources of information):

- tool description,
- who uses the tool,
- cost,
- strengths,
- limitations,
- complexity,
- practical application,
- typical implementation period,
- references,
- key words, and
- Internet Uniform Resource Locators (URLs).

(As you are no doubt aware, the Internet is a dynamic entity that can change daily. We apologize in advance if any listed URL address no longer works for you. Downloading files and data from websites can be risky. Virus-scanning software can help avoid problems. However, we urge caution and assume no responsibility for website content or download problems encountered due to visits to the sites herein recommended.)

GROWTH DREAMS . . . AND NIGHTMARES

Growing up is great, but it can also be stressful and tough at times: the first day of school, first date, first year away from home, getting married, first child, etc. Growing a business also has many firsts. As companies grow, much like people, they face stepping stones, milestones, and stumbling stones.

To avoid stumbling along on their own in their lives, many people see the value of mentorship—partnering with another individual

Table 1-2. Troubleshooting guide

Find the problem you are experiencing and see if there are any "●s" beneath it. If so, look across to the tools listed and then go to that section of the book that discusses the tool.

		Problems													
		No formal plan to improve	No quality system structure	No structured problem solving	Process variability	Unacceptable quality/reject rate	Unclear customer expectations	Vague process knowledge	Ineffective teams and projects	Decisions made on opinion	Unacceptable lead times	Unacceptable inventory levels	No sense of team	Poor supplier performance	Excessive costs
Higher order	Six Sigma	●	●	●	●	●	●	●	●	●	●	●	●	●	●
	Baldrige	●	●		●	●	●								
	Deming Prize	●	●		●	●	●								
	ISO 9000	●	●		●	●	●								
	Total Quality Management (TQM)	●		●					●	●			●		
	Total Productive Maintenance (TPM)	●			●	●					●	●			●
Define	Suppliers, inputs, processes, outputs, and customers (SIPOC)			●	●	●	●	●	●	●			●		
	Quality Function Deployment (QFD)			●			●			●					
	Critical to Quality (CTQ) Tree			●	●	●	●		●	●				●	
	Voice of the Customer (VOC)						●		●	●				●	
	Hoshin Kanri (policy deployment)	●								●			●		
	Affinity diagram			●		●		●	●				●		
	Product, quantity, routing, support systems, and time (PQRST)							●		●					

Table 1-2. (continued)

		Problems													
		No formal plan to improve	No quality system structure	No structured problem solving	Process variability	Unacceptable quality/reject rate	Unclear customer expectations	Vague process knowledge	Ineffective teams and projects	Decisions made on opinion	Unacceptable lead times	Unacceptable inventory levels	No sense of team	Poor supplier performance	Excessive costs
Measure	Statistical techniques			●	●	●		●	●	●				●	
	Spaghetti diagram							●			●	●			●
	Value Stream Mapping			●				●	●	●	●	●			●
	The dashboard	●	●		●			●		●					●
Analyze	5-Whys			●	●	●		●		●	●	●		●	●
	Design of Experiments (DOE)			●	●	●		●	●	●				●	●
	TRIZ (theory of inventive problem solving)				●	●									
	Quality mapping			●	●	●		●		●					

Table 1-2. (continued)

		Problems													
		No formal plan to improve	No quality system structure	No structured problem solving	Process variability	Unacceptable quality/reject rate	Unclear customer expectations	Vague process knowledge	Ineffective teams and projects	Decisions made on opinion	Unacceptable lead times	Unacceptable inventory levels	No sense of team	Poor supplier performance	Excessive costs
Improve	Failure Mode and Effects Analysis (FMEA)			●	●	●									
	Kaizen	●		●	●				●	●	●	●	●	●	●
	5-S (sort, systematize, shine, standardize, and sustain)	●			●								●		●
	Set-up reduction (Single-Minute Exchange of Dies [SMED])				●	●					●	●			●
	Flow										●	●			●
	Kanban										●	●			●
	Cellular configuration								●	●	●	●	●		●
	Team development			●		●		●	●	●			●		●
	Corrective action	●	●	●	●	●		●		●					●
Control	Takt time										●	●			●
	Line balancing										●	●	●		●
	Poka-Yoke/Jidoka	●		●	●	●									●
	Best practices				●	●	●	●		●	●	●			●
Other	Supply chain management	●			●	●		●		●	●	●		●	●

to seek wisdom through shared knowledge and experiences, rather than trying to gain all wisdom first hand through trial and error. During the period of maturation associated with transforming a company from business as usual to one capable of operating at Six-Sigma performance levels, you should expect a certain amount of stress, but there should be no need to feel you are out there all alone. You can go to trade shows and seminars. You can talk with other companies who are also on this path. You may find a company or a person with whom you can partner—to learn with, share ideas with, and either mentor or be mentored. There is as much to gain for the mentor as there is for the one being mentored: sharing visits to each other's companies, talking about successes and failures, and benefiting from cross pollination between your project teams and their project teams as you share Kaizen team members.

MATCHING TOOLS TO OBJECTIVES

Like other shade-tree mechanics, I enjoy going to my local home improvement store and discovering a new tool I can use—a lighted screwdriver or a multi-function wrench, or a magnetic gadget for picking up dropped nuts or bolts. I love adding such new tools to my collection, and I sometimes deliberately leave them out for a while so visitors can ask me about them. Yet, sooner or later these new tools become just another tool in my toolbox like everything else.

Some companies approach new World-Class techniques in a similar way. Every few years another tool is introduced (or rediscovered) and because of a particular success at some renowned company, everyone drops what they are doing and runs out to buy a book or learn about this new improvement device. In the process, initiatives that everyone had been focusing on become old news and previous projects are abandoned in mid-stream. To quote cartoon character Ziggy, "Today in the newspaper, tomorrow in the bird cage."

My message here is not to disparage such improvement, and I certainly do not wish to discourage anyone from seeking to become World Class. My point is that a lot of these tools have been discovered and many may not be new at all. They may simply be refinements of tools already in your toolbox. In their efforts to keep the improvement message fresh and energy levels high, managers often get enamored with these new tools, not realizing the damage they can do by uprooting countless hours of team effort spent to implement the "last" next new tool.

Just because I buy a new screwdriver that has interchangeable bits, does not mean that I will go home and throw out all my other screwdrivers. Similarly, the discovery of new continuous-improvement techniques should not result in abandoning a company's vision every few years. Tools may be introduced or improved every so often, but the vision should not change.

What should that vision be? Regardless of how the company phrases it, it should be *to be the best at what it does,* regardless of the company's name, size, industry, or competitors. Each company should strive to be World Class. That basic vision should be unchangeable.

WORLD CLASS DEFINED

The term World Class is often applied to athletics. Think about some World-Class athletes like Michael Jordan, Larry Bird, and Tiger Woods. What are some of the characteristics and behaviors that set these athletes apart as World Class?

- dedication,
- flexibility,
- training,
- motivation,
- talent,
- willingness to do whatever it takes,
- creativity,
- citizenship (role model),
- determination, and
- willingness to adapt.

Now, think about some of the companies in the world who have set themselves apart from the ordinary companies: companies like Kodak, General Electric, Harley Davidson, and Toyota. What are some of the characteristics and behaviors that set these organizations apart as World Class from the endless list of ordinary organizations? It's the same list isn't it?

Take another look at the list of characteristics. It is really more related to software (people relationships) rather than hardware (machines, buildings, materials, etc.). If a company hopes to compete—and set the pace for the competition—then 100% of the company's team needs to operate as World-Class competitors. Otherwise the company will simply be ordinary.

EVERY DAY A NEW SCORE

One hard reality is that a company's customers have short memories when it comes to performance. Think for a moment about your favorite sports team and its love-hate relationship with its fans. When the team is on a winning streak, it can do no wrong in the eyes of its adoring fans. Slip into a losing streak, however, and those fans attack it like starving piranha. This same dynamic occurs in the vendor-customer relationship more often than is realized. There's an old saying, "One happy customer will tell three people. One unhappy customer will tell 20." It is expensive to dissatisfy a customer. Like Dr. Deming would say, it is immeasurable.

The nice thing about sports is that no matter how badly a team did yesterday, today's game starts with the score 0 to 0. Sometimes people need to be reminded that each day is a fresh new start. To a customer, a company is only as good as it performs today. Yesterday has been forgotten. The scoreboard reads 0 to 0 every morning, and the company is in competition with other suppliers who will try to outperform it today. The team that best satisfies that customer today wins the game. Being World Class includes avoiding any opportunity for the customer to be dissatisfied.

How can a company hope to accomplish this? Think about your water heater for a moment. The heating element goes out, so off you go to the local hardware store and find a $29 element in stock, but note that its quality is seriously in question because it carries no Underwriters' Laboratories (UL) listing. So you decide to continue your search. On your second stop, you find a salesperson who offers a $39 version that is UL listed, but it is not currently in stock and must be ordered for pickup in four days. Since this delay is unacceptable, you seek out yet another store where you find a $49 fully tested element that is available. Grudgingly, you pay the extra price.

What if that first store had a $29 version that met your quality expectations and was in stock? It would be a no-brainer decision. This is exactly the condition a company needs to create with its customers. If the quality-cost-delivery criteria that most savvy customers demand today can be met or exceeded, then a no-brainer situation has been created. Having customers realize that they can't go anywhere else to get the quality, service, reliability, dependability, availability, affordability, and value your company offers ensures you are on the right side of the zero-sum equation. This kind of performance is what it means to be World Class—to perform like Michael Jordan or Tiger Woods. No other company really has a chance to catch up if your company is at the top of its game. In the words of one company vision statement: "We intend to be the standard by which all competitors are measured."

To be World Class, all aspects of the business must be well managed. Six Sigma, Lean Manufacturing, Just-In-Time (JIT), Total Quality Management (TQM), and all the other tools can not be applied simply and solely within the manufacturing environment. Customers need to see quality processes from the first stage of phone communication, through order entry, design, planning, manufacture, delivery, and service. Each activity must be brought under the bright light of scrutiny.

SOMETHING FISHY HERE?

No activity in the organization can be considered sacred or out of bounds from the potential application of continuous improvement tools. A simple cause-and-effect diagram normally used to identify potential root causes of problems can be used to describe how every aspect and activity in a company needs to be analyzed to yield the greatest potential for improvement. Figure 1-3 is a fishbone or Ishikawa diagram that shows the desired effect of becoming World Class. The inputs or causes to this effect are shown at the end of each branch of the fishbone. In this example, all inputs start with an M, just as a memory aid.

By adding a few sample causes to the fishbone example in Figure 1-3, it can be seen how these activities can either positively or negatively affect the outcome, as shown in Figure 1-4. It soon becomes apparent that no organizational activity can be overlooked

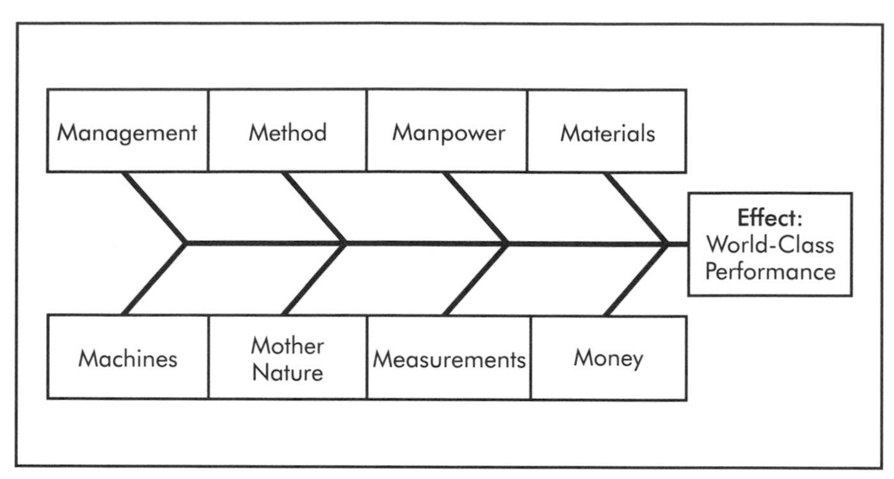

Figure 1-3. Ishikawa cause-and-effect diagram (fishbone).

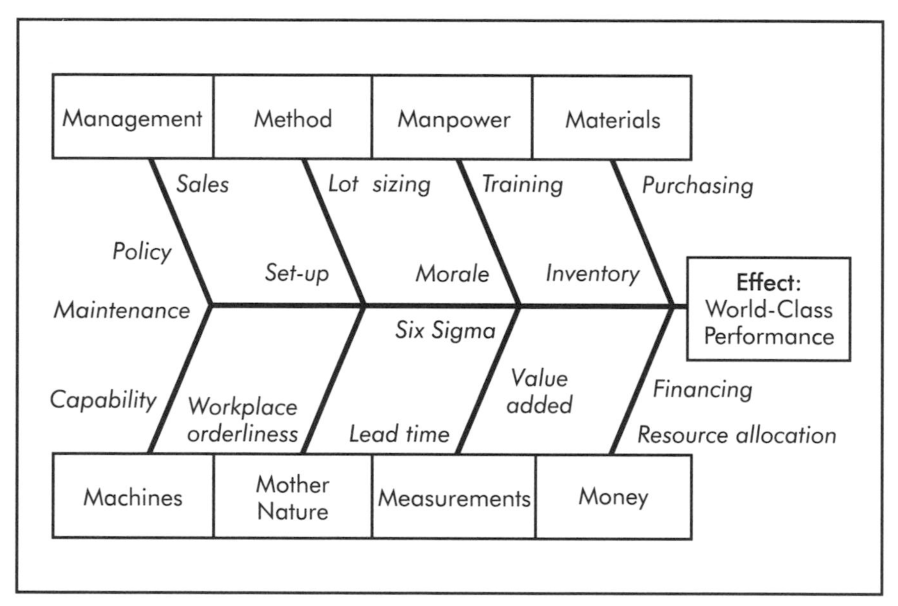

Figure 1-4. Sample cause-and-effect diagram.

if a system that approaches a level of Six-Sigma performance is to be realized. This is by no means a complete fishbone diagram, but it begins to show the many inputs there are within even a small company.

SORTING THROUGH THE TOOLBOX

With all the causes in Figure 1-4 to manage (and more), it's no wonder that over the years many tools have been developed to help measure, control, and improve the individual components of a business. Management must be constantly aware of the risk of over-controlling (like over-steering on ice). If not careful, one aspect of the business could be optimized at the expense of suboptimizing the whole. This is where the holistic approach and Six Sigma comes in: looking at all the tools as having to dovetail together into a seamless process of improvement, where one improvement does not negate the improvement made yesterday or yesteryear. This requires measurement— and conscious and constant diligence.

Don't let the number of tools throw you. You do not need to learn every tool at once. In this book, each tool is introduced to you as it makes sense. One of the goals is to eliminate the feeling that you must juggle all these tools at once (see Figure 1-5).

Figure 1-5. Juggling all the options.

Figure 1-6 outlines some common and primary tools used by World-Class companies. Within these primary tools are subsets of tools that will be discussed in later chapters. Entire books have been written about many of these individual topics. The intent is not to duplicate this library of work. An overview of each tool in the toolbox precedes a list of additional resource materials and websites where you can find more information. The goal is to provide a sense of order so you can see clearly how each tool is categorized and where its use might be most appropriate.

Define	Measure	Analyze	Improve	Control
Voice of the Customer (VOC)	Histogram	X and R charts	Single-Minute Exchange of Dies (SMED)	Best Practice (SOP)
Affinity diagram	Value Stream Mapping	TRIZ	Kaizen	Visuals
Critical to Quality (CTQ)	Failure Mode and Effects Analysis (FMEA)	Design of Experiments (DOE)	Quality circles	Jidoka
Product, quantity, routing, support systems, and time (PQRST)	Supply Chain Management	Management By Walking Around (MBWA)	FMEA	Poka-Yoke
Suppliers, inputs processes, outputs, and customers (SIPOC)	Hoshin Kanri	Ishikawa	5-Whys	Andon

Figure 1-6. How various tools fit under the Six-Sigma umbrella.

REFERENCE

Conner, Gary. 2001. *Lean Manufactruing for the Small Shop.* Dearborn, MI: Society of Manufacturing Engineers.

2

Working with Six Sigma

In this chapter, the fundamental concepts of Six Sigma are introduced, which includes looking at product or process variation and how to make this variability visible and controllable. After explaining the basic math behind Six Sigma, a spreadsheet is designed to help you identify process variability. The use of the define, measure, analyze, improve, and control (DMAIC) formula is explored along with how Six Sigma (or any World-Class initiative) might be organized within your company.

The term Six Sigma is unique in that it has become recognized as both a measurable objective or goal and an expression used to describe the many tools employed to attain that objective. It is an approach used to improve any process, the ultimate objective being to control variation and waste to improve profitability, quality, cost, delivery, and reliability.

The aggregate benefits to the hundreds of companies applying Six Sigma today can be measured in billions of dollars. Not everyone has embraced the approach, of course, but the evidence is in, and the process has proven itself in nearly every industry, regardless of company size.

PROCESS VARIATION—NOT SO GOOD?

The Six-Sigma concept is based on the fact that all processes vary to some degree. Variation can be a good thing in some cases. For example, what if automobile headlights all burned out exactly at the same time? Variation here means headlights fail one at a time, and this allows us to limp home with at least partial light—much better than having both headlights go out at once.

Generally, though, variation is seldom beneficial and not something that should be built in or allowed in manufacturing or service processes. Think again about your car. If you need to replace a component, any variation in the manufacturing process could result in the new part not fitting correctly. The color may not match, or the finish may look or even feel different. Variation can be subtle, yes, but if excess variation is allowed, units may be produced that do not function at all.

So variation is not good and critical parameters in the manufacturing process must be measured if a company is to bring them under control. Measuring and calculating variation from the desired norm and then plotting the data yields what is commonly called a bell-shaped curve, as shown in Figure 2-1. It is also known as a Gaussian curve after the German mathematician who defined its theoretical properties.

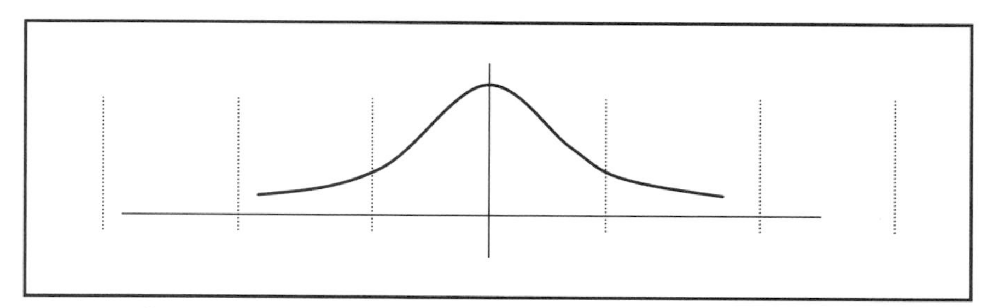

Figure 2-1. Simple bell-curve distribution.

If a simple set of lines known as *sigma limits* are overlayed onto this graph, you can begin to see (and therefore begin to control) how a process is performing against the goal of producing perfect parts every time. This is much like trying to learn to drive by staying between the lines on the road.

GOOD SIGMA VERSUS BAD

A normal bell-shape curve theoretically represents 100% of the parts being measured. The curvature is smooth and symmetrical, and its center or peak represents the most common or average value, and the extremes, left and right, the measurements that are the furthest "off center." If the bell-shape curve is not normal (for example, if there are two or more peaks, or if the shape is very irregular) there may be a special cause for the variation, making the use of Six-Sigma tools less effective until the cause(s) is eliminated. In other words, the process has to be in a relative state of control before even trying to adjust it using Six-Sigma techniques. Otherwise, it's like driving on ice, you have little control or tend to over-control (over-steer).

Adding Sigma Limits

By dividing the scale into six sections or limits of deviation from the center value, three on each side of the target, a level of confidence can be determined for how consistently the product measurement can be held to the required specification over time. These divisions, or sigma limits, are calculated in the following section.

It can be shown mathematically that with a normal distribution of measurements, you can expect to see approximately 68% of all the measurements fall within the first sigma limit on each side of the target, as shown in Figure 2-2.

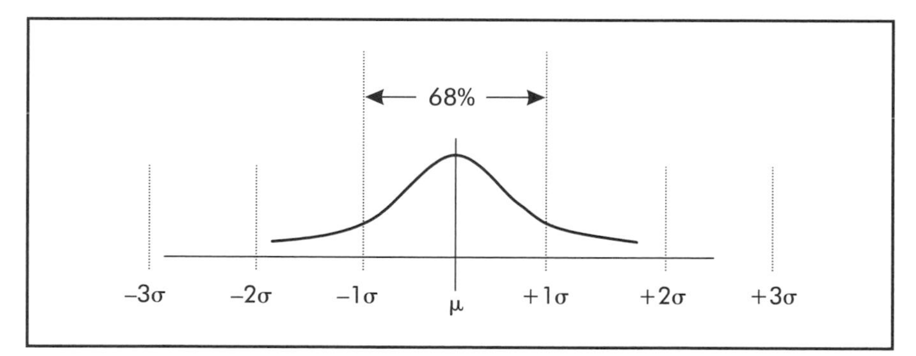

Figure 2-2. Bell-curve distribution with 68% population within 1-sigma limit.

If a greater percentage of the parts or process being measured fall closer to the target, then the process has demonstrated it is more capable of producing parts with less variation. The sigma limits would then be adjusted to represent the improvement shown in Figure 2-3.

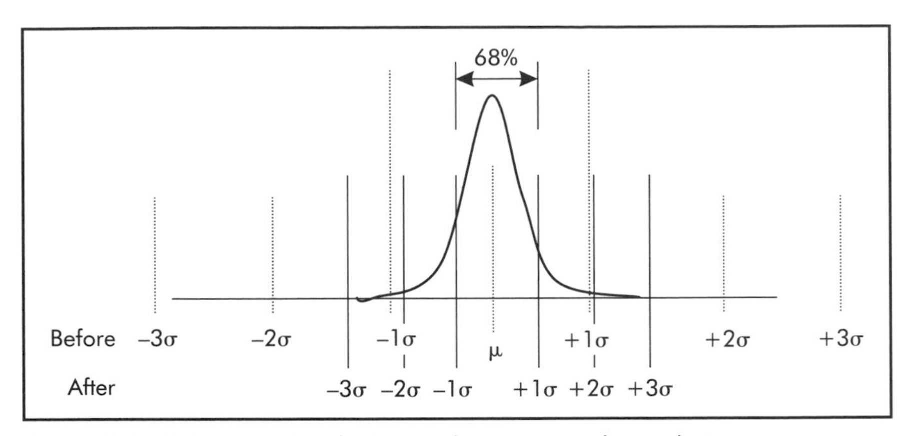

Figure 2-3. Bell-curve distribution with a narrowed population.

If all parts fall under a normal bell-shaped curve, and within three sigma limits (and within customer specifications) on a Six-Sigma grid as shown in Figure 2-4, (+3σ and –3σ), then theoretically only 3.4 defects would be produced for every million opportunities. The process is then said to be operating at Six-Sigma performance levels.

The goal is to have a narrow base that falls within +3 and –3 sigma limits. If any parts fall outside of these limits, there is a chance that unacceptable variation (a defect as defined by the customer) will be produced.

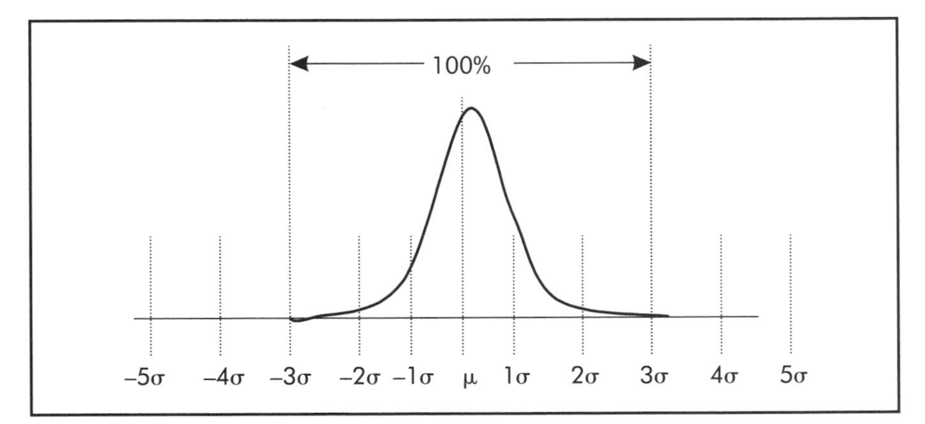

Figure 2-4. Six-sigma chart with 100% of variation within the three-sigma limit.

To maintain variation within acceptable limits, the process must be strictly controlled. If the deviation is large, then the process clearly is producing products that vary a great deal and many will fall outside specifications. Used to identify how large the deviations are, sigma is simply a statistical measuring tool.

If something is 60 miles (96.5 km) away, you know that it is much farther than 10 miles (16 km) away. For example, if a process has a sigma of 0.600 in. (15.24 mm), then you know the variation is much larger than a process with a sigma of 0.100 in. (2.54 mm), which indicates the former process is not as well controlled as the latter. The objective behind Six Sigma is to make sure that sigma—the standard deviation of a process—is as small (narrow) as possible, indicating it allows the least amount of variation.

Calculating Sigma Limits

In actual practice, a spreadsheet can be used to do the math, but it is important first to understand how the sigma limit is calculated. The basic equation is:

$$\sigma = \sqrt{[\Sigma(x - \bar{X})^2] \div (n - 1)} \tag{2-1}$$

where:

σ = one standard deviation (one sigma)
x = each individual dimension from the sample, in. (mm)
$\bar{X}$ = average of all dimensions, in. (mm)
n = number of parts in the sample

Imagine you own a small manufacturing company that makes cement building blocks. Each block is supposed to weigh 20 lb (9 kg), but due to variation in the amount of water, drying time, sand mixture, etc., some of the blocks weigh more and some weigh less. The blocks can weigh as much as 27 lb (12 kg) and as little as 13 lb (6 kg). If they are greater than the upper limit of 25 lb (11 kg) or below the lower limit of 15 lb (7 kg), then they are considered unacceptable and scrapped (crushed).

So, to find the sigma for this process, five cement blocks are weighed. Here:

n = 5
x_1 = 21.0

x_2 = 22.0
x_3 = 19.0
x_4 = 20.0
x_5 = 19.0

$$\bar{X} = (x_1 + x_2 + x_3 + x_4 + x_5) \div 5 = 20.2$$

where:

x_1, x_2, x_3, x_4, x_5 = weight of each part in the sample

So:

$$\sigma = \sqrt{[(21-20.2)^2 + (22-20.2)^2 + (19-20.2)^2 + (20-20.2)^2 + (19-20.2)^2] \div 4}$$

$$\sigma = \sqrt{[(0.8)^2 + (1.8)^2 + (-1.2)^2 + (-0.2)^2 + (-1.2)^2] \div 4}$$

$$\sigma = \sqrt{[(0.64) + (3.24) + (1.44) + (0.04) + (1.44)] \div 4}$$

(negative variances squared are positives)

$$\sigma = \sqrt{[(0.64) + (3.24) + (1.44) + (0.04) + (1.44)] \div 4}$$

$$\sigma = \sqrt{6.8 \div 4}$$

$$\sigma = \sqrt{1.7}$$

$$\sigma = 1.304$$

Thus, the sigma value for this example is 1.304. This value is better than a sigma of 4.304, but not as good as a sigma of 0.304. (Note that in actual practice more samples are taken to more accurately measure the capability of the process.)

Zeroing in on Six Sigma

If 100 of the cement building blocks are stacked according to weight (assuming there is a stable process), you might end up with a stack like the one shown in Figure 2-5 (approximating a bell-shaped curve). There is obviously a problem here. There are parts that fall outside the acceptable limits, and more that are very close to being rejected. To keep the customer satisfied (and avoid throwing so many parts away), the reason for this variation

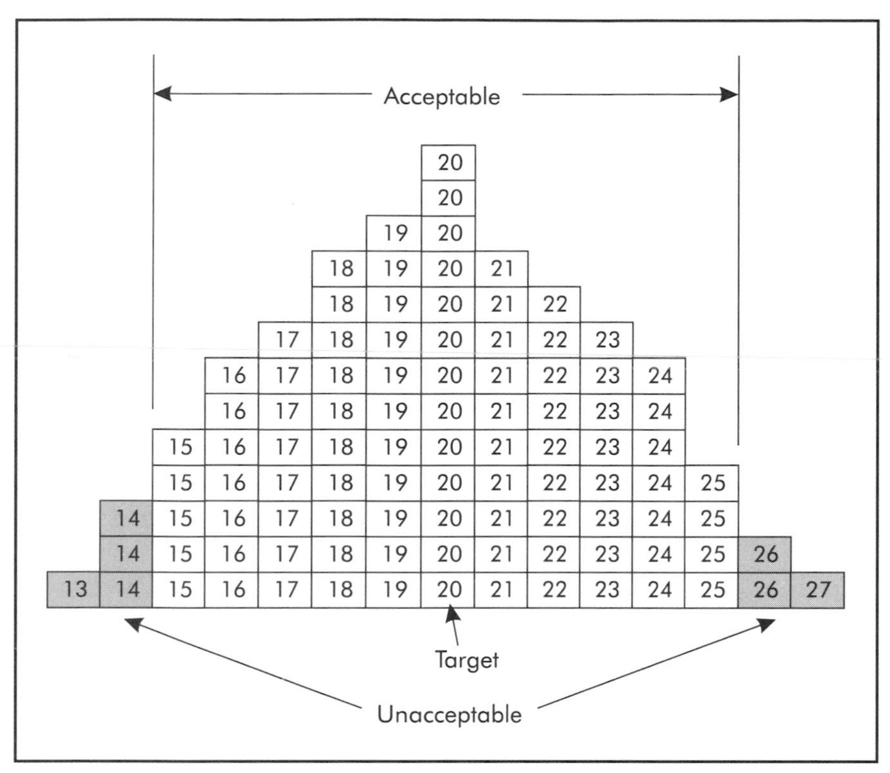

Figure 2-5. Distribution of cement blocks by weight.

needs to be identified and controlled. The tools of Six Sigma (and other tools described later) will help accomplish this goal.

The sigma-limit calculation for the block-building process is shown in Table 2-1. (Note: There are other factors shown here that will be discussed later.)

The resulting sigma calculation as shown in Table 2-1 is 3.166. If this number is used as a dividing line, overlaying three sigma limits (3×9.497) in each direction, the third sigma reaches far beyond the customer specification. The goal is to have the Six-Sigma limits fall inside the customer specification of ± 5.000. From this, you can get a picture of how close to Six Sigma the process is performing. (Not too well in this case. If you were driving a car, and the customer specification was the fog lines on the side of the road, you would be in the ditch!)

Table 2-1. Sigma limit calculator

Target	20.000
Upper tolerance	5.000
Lower tolerance	5.000
Upper specification limit (USL)	25.000
Lower specification limit (LSL)	15.000
Average	19.910
Sigma	3.166
3 Sigma	9.497
6 Sigma	18.995
High	27.000
Low	13.000
Range	14.000
C_{pk}	0.517
C_p	0.526

If the part measurements and the Six-Sigma limits fall within the acceptable limits set by the customer, then everyone is happy. If on the other hand, the parts fall outside the three-sigma dimension, then the process is operating at less than Six Sigma and the chances increase that sooner or later a defect will be created. In this example, three-sigma limits equal ±9.497 in both directions from the target. The parts are falling within the third sigma limit (±7.000) on each side of the target, but far outside what the customer specifies as acceptable. Thus, the defective parts will need to be sorted.

In Figure 2-6, the third sigma limit is far outside the acceptable-variation limits on both sides of the bell curve. This is not an acceptable situation, and the reasons for this variation must be found and controls put into place to prohibit variation from exceeding the established tolerance.

In the next example, Figure 2-7, the process looks much improved—none of the parts fall outside acceptable limits. Yet the process is still not operating at a Six-Sigma level of performance. Additional measures need to be taken to ensure no parts fall outside the acceptable limits over time. Although there is substantial improvement in performance and sigma value (from 3.166 to 1.502) as shown in Table 2-2, more work needs to be done to be operating at Six Sigma.

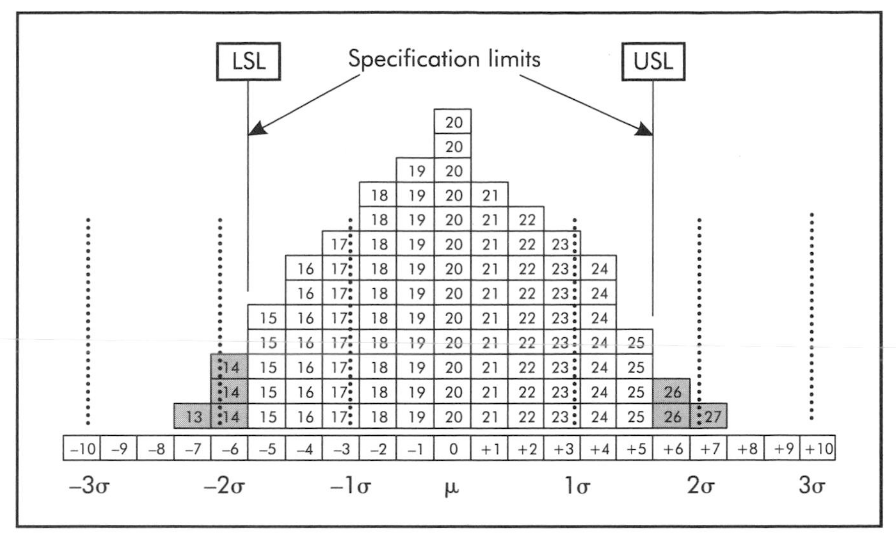

Figure 2-6. Cement-block distribution on a three-sigma grid.

To insure a defect rate of less than 3.4 per million opportunities—to be operating at Six Sigma—the process must generate a sigma limit somewhere around 1.25. This would require a normal distribution of parts all falling within 3.75 on either side of the target. Even though the customer would accept parts slightly outside that range, the process must have even tighter control to ensure no part is likely to become a quality suspect.

PROCESS IMPROVEMENT THE SIX-SIGMA WAY

The define, measure, analyze, improve, and control (DMAIC) process is key to improvement. Begin by defining the current condition (the problem) and the variables involved. For example, the variables in the concrete-block manufacturing process include:

- water (amount and temperature),
- concrete (amount, temperature, moisture content),
- sand (amount, temperature, moisture content),
- mixing time,
- mixing speed,
- ingredient-addition sequence,
- operator performance (consistency, level of training),

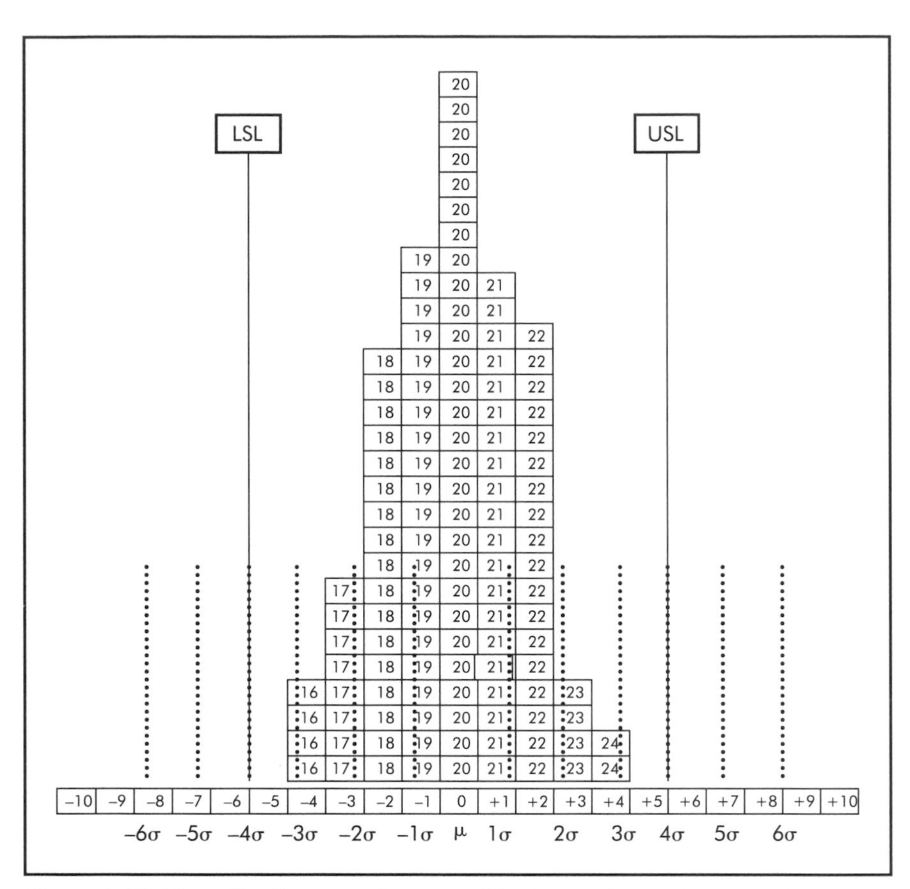

Figure 2-7. New distribution of cement blocks on three-sigma grid.

- ambient temperature,
- humidity, and
- machine cleanliness.

In this case, the method used by the operator to add water, concrete, and sand may vary. The additions made by each operator may vary day to day or operator to operator. The recipe may be subjective, with each operator using slightly different amounts. The percentage of each ingredient needs to be calculated and controlled based on the size of the batch, and there also may be a need to modify the recipe based on different weather conditions. All of these variations need to be addressed to adequately define the current condition.

Table 2-2. Improved result according to sigma limit calculator

Target	20.000
Upper tolerance	5.000
Lower tolerance	5.000
Upper specification limit (USL)	25.000
Lower specification limit (LSL)	15.000
Average	19.870
Sigma	1.502
3 Sigma	4.506
6 Sigma	9.012
High	23.000
Low	16.000
Range	7.000
C_{pk}	1.081
C_p	1.110

Next, the measuring step provides real data upon which to base decisions. The collection and graphing of such critical process variables as block weight or mixture consistency at different points in the pour cycle might reveal hidden reasons for variation. Measuring critical characteristics over time can also help identify process creep—problems that show up over long periods of time—due to machine wear, different material suppliers, changing formulas, or new operators.

You do not need to have a degree in mathematics or statistical process control (SPC) to use measurements to help analyze a process. There are excellent reference books to help simplify these tools for you (some are listed in the bibliography of this book).

The analyzing step is often overlooked in favor of jumping directly to a solution. This is a challenge particularly for managers who often see their roles as decision-makers. The analyzing step usually includes breaking down the process being examined into discrete steps. A process flow diagram, examination of the operator's standard work process, or a quality map can often help identify where variability is being introduced. It is important to include the operator (and his/her input) in the analysis process.

The improving step is the fun part of the process because here people get a chance to be creative. Creativity is exactly what is needed, not just creative ideas for controlling variables, but ideas for making the job easier, safer, less stressful, less dependant on someone remembering a key step, more automatic, more productive, more repeatable, and more likely to provide the customer a satisfactory product or service. Again, the operator and value-stream owner (manager of the process) should be involved. This step also includes finding a way to test and measure the results of the many ideas generated.

The last step is controlling. Improvements have a very short half-life if sustaining them is left to chance. Putting measures (figuratively and literally) into place that require behaviors to be modified will help to improve sustainability. Some assembly processes use overlay templates and require personnel to touch each screw or nut with their fingers and count rather than depend on a quick inspection glance. This can save hours or even days of field repair work on a computer or telecommunication housing. Missing nuts and screws found later when circuit boards are field installed may require tearing down an entire assembly.

Many More Opportunities

Think about all the aspects of the cement-block manufacturing business where this methodology could be applied. The accuracy of how orders are taken also can be measured. How many potential errors can be made while performing some of the following routine activities? Do you know the defects per million opportunities (DPMO) associated with the following activities in your operation?

- purchasing,
- order entry,
- work-order development,
- scheduling,
- finance,
- sales,
- production/manufacturing,
- production planning,

- production engineering,
- machine operation,
- new-product introduction,
- human-resource management,
- maintenance,
- industrial and manufacturing engineering, and
- traffic and delivery systems?

Obviously, the five DMAIC steps and the process of improvement using Six Sigma work for more than just manufacturing cement blocks. Here are a few examples in other industries that could benefit from this same process:

- on-time departures (airlines);
- surgery recovery time;
- on-time delivery;
- machine capability (machine-tool manufacturers);
- restaurant food quality (taste, timeliness, temperature, etc.);
- hotel-room preparedness;
- clinic-patient wait time;
- insurance-claim process time and accuracy; and
- phone-call reception (timeliness, accuracy, helpfulness, etc.).

Finding the Right Tools

Once an opportunity (or problem) has been identified, there are countless methods that can be used to gather measurements, analyze, improve, and control the condition you are trying to enhance. The remaining tools described in this book are not meant to identify them all, or to replace a pure Six-Sigma approach. The goal is simply to expose you to some of the more commonly used tools and enhance your ability to use the five-step DMAIC process. You will find many opportunities to use these techniques to narrow down the variability in critical processes and consistently meet the needs of your customer.

By defining, measuring, analyzing, improving, and controlling the right things, it should then be possible to roll up the individual performance measurements into one *aggregate* value or number that will indicate the overall synergistic improvements (and health) of the organization (see Figure 2-8).

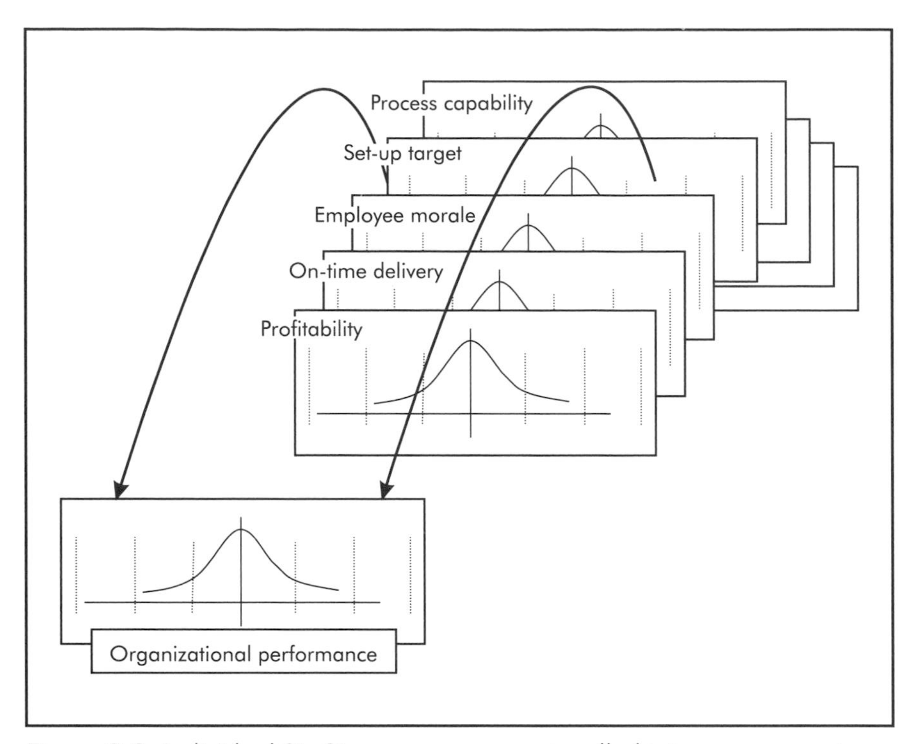

Figure 2-8. Individual Six-Sigma measurements rolled into an aggregate organizational metric.

As with most initiatives for change, Six Sigma and other similar tools have much overlap, with many of the terms and techniques common to one another. Instead of describing each major tool in great detail (for example, Lean Manufacturing, JIT, TQM, etc.) and duplicating the sub-tools within each methodology, this book offers an overview at the macro level (higher-order tools) and then defines as many of the lower-order support tools as possible. Many of these tools are used not only by champions of Six Sigma, but also within Lean Manufacturing, TQM, JIT, and other methodologies. Readers will see and determine how and when each of these tools can best support their efforts at the opportune time (see Table 1-2). Each tool is described in terms of where it fits in the DMAIC cycle, the costs associated with the tool, its strengths and limitations, how complex it is to apply, and wherever possible, examples of practical application.

ROLES IN THE SIX-SIGMA ORGANIZATION

There is no "one-size-fits-all" program, and setting up a Six-Sigma structure within your company is just the beginning. Figure 2-9 shows how companies can approach the duties required to carry out a Six Sigma or Lean transformation. It shows an organizational chart that maps the chain of command and activities performed by each position. Training requirements for these positions are discussed later in this book.

Depending on the size of a company, there may be only one person filling the role of champion, black belt, and green belt as well as performing other duties. The champion is the ultimate change agent. His or her role should be defined as the one taking the lead in Six-Sigma implementation. This would include setting the overall strategy for implementation, performing justification calculations for projects, and obtaining resources, including people, funding, and assistance from value-stream managers and support functions.

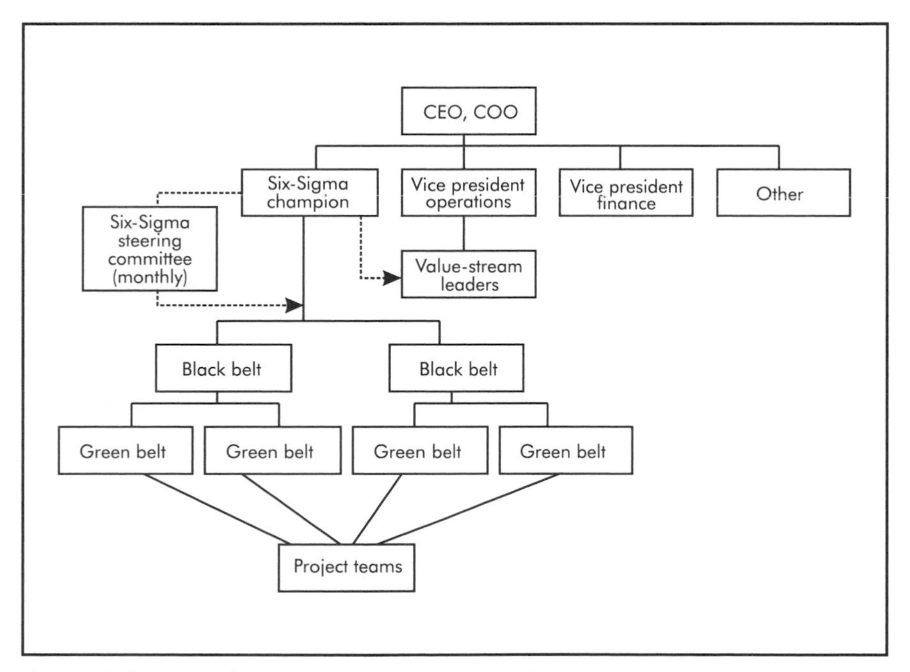

Figure 2-9. Typical Six-Sigma organizational structure.

The champion also may be the standing chairperson for the Six-Sigma steering team, generally made up of senior staff. This group would meet monthly and help select projects based on team recommendations, or the data and information gathered by the black belts and champion. The steering team would also attend all team update presentations.

The champion should participate in the selection of master black belts and black belts along with developing the criteria by which they are selected. He or she should develop a Six-Sigma training program. Other responsibilities would include approving and auditing projects and reporting to the steering team on implementation success.

The black belt's role is to assist project teams, and to teach the green belts how to become black belts. Black belts are not necessarily full-time team members. They are called in when needed, and participate in the daily wrap-up meetings to offer guidance along with the steering committee.

Green belts are generally part-time project-team members who have participated in more than one project. They can help new people by showing them the process and how it works. After having participated in a number of projects, green belts begin to take on a more meaningful role, including leading a project team.

Table 2-3 summarizes the Six-Sigma tool.

Table 2-3. Six-Sigma tool summary

Who needs and uses Six Sigma	Companies desiring to set themselves apart as World Class rather than ordinary
Cost	High, depending on current conditions
Strengths	Promotes a systematic approach to eliminating waste, not trial and error—the Six-Sigma approach is time tested, not new.
Limitations	Six Sigma requires trained, dedicated resources, and sustained effort. It will not run itself.
Process complexity	Moderate to difficult
Implementation time	2–10 years
Additional resources	See Bibliography
Internet search key words	Six Sigma, process improvement, quality initiatives
Internet URLs	www.sme.org/sixsigma www.processassociates.com/bookshelf/subjects/s_30_5.htm www.isixsigma.com/ www.rathstrong.com www.statsoft.com/textbook/stathome.html www.analyse-it.com/default.asp

3

Higher-order Tools

This chapter focuses on sorting out some of the major continuous improvement tools offered. The discussion presents an overview only. For detailed coverage of these tools, see the bibliography.

This chapter will first discuss the Toyota Production System and how the world of manufacturing continues to learn from Toyota's Just-In-Time transformation in the 1970s and 1980s. Then, you will see how Group Technology can help a team see the individual "trees within the forest" and make opportunities for improvement visible and more likely. Total Quality Control and Total Quality Management tools also will be examined, along with a short review of the Theory of Constraints. The importance of equipment care is highlighted in the section dealing with Total Productive Maintenance.

Each tool has elements that dovetail with the Six-Sigma approach, and the objective here is to illustrate some of the similarities so that the synergy is more obvious than the differences. Finally, this chapter reviews some of the more popular quality initiatives from the last few decades including ISO 9000, and the Baldrige and Deming approaches. The objective is to reinforce the idea that all these tools harmonize with a company's goal of being World Class.

TOYOTA PRODUCTION SYSTEM

The Toyota Production System (TPS) and other organizational models like it (for example, the Ford Production System [FPS]) were developed by organizations to approach continuous-improvement (CI) efforts more holistically. Primarily developed and evolved at Toyota to bring a level of competitiveness to Japan's post-war economy, the Toyota Production System is often held up as an ideal production system, especially for suppliers in the automobile industry.

The mantra: "Takt time, one-piece flow, and pull systems" has become the theme song at many US and European companies trying to match Toyota's level of improvements in throughput, productivity, inventory reduction, cost improvement, set-up reduction, and workplace organization.

The Toyota Production System is often described as a house built upon a solid foundation. Toyota identified a sound quality system as the basic underpinning to support the rest of TPS. (In this, TPS and Six Sigma agree strongly.) Total Quality Management must provide the substructure. Upon this foundation, a layer is added called "production smoothing," which includes developing a comprehensive sales, inventory, and production-planning system. Next, the two main pillars (principles or tools), Just-In-Time (JIT) and Jidoka, give the structure its strength and durability. Figure 3-1 shows how this system is conceptualized for people inside the organization as well as vendors, customers, and those who visit Toyota.

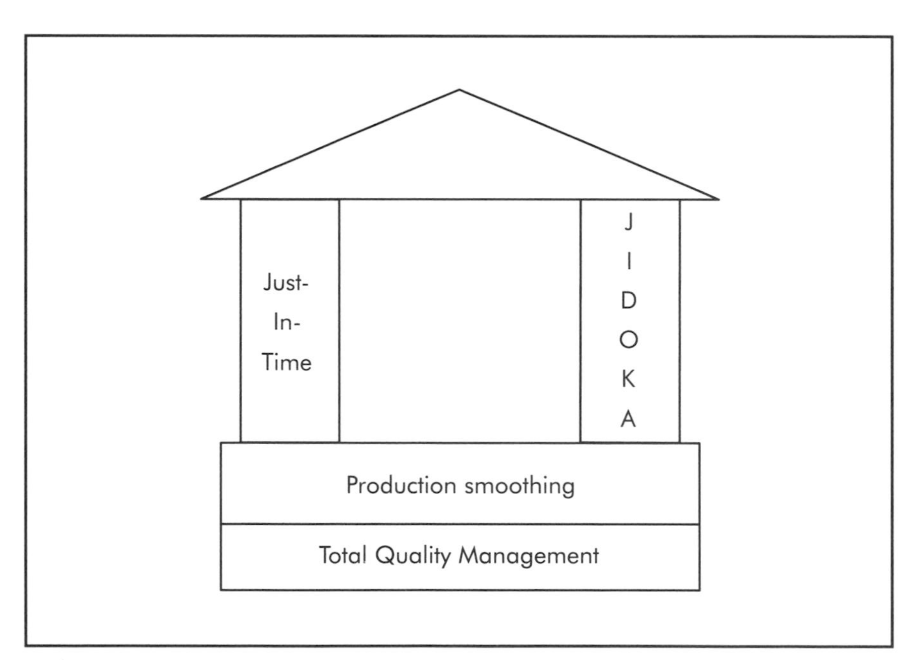

Figure 3-1. Key elements of the Toyota Production System.

Many of the JIT techniques—which include techniques like set-up reduction, pull systems, and Jidoka—are tools that can be applied to a small company. However, for a small company to try and act like Toyota is much the same as trying to pretend to be a General Electric in terms of applying Six Sigma. The shoe will not fit. A hybrid approach must be taken to retain the flexibility and nimble nature required by a make-to-order or an engineer-to-order manufacturer. This is not to say that your company cannot benefit from the processes employed at Toyota or Ford. It is just a little harder and takes a good deal more creativity to apply the tools to the job shop. If, for example, you have 3,500 active part numbers—all distinct in size, shape, and even material type—and you never know which ones you will sell tomorrow, then you obviously have a level of complexity greater than many larger organizations. Production smoothing works best when you can predict with some confidence what you will sell tomorrow or next week. Job shops often do not have that luxury.

There is never one tool to fix all the problems. However, the Toyota Production System, as shown in Table 3-1, does contain many of the same tools used in the Six Sigma approach. By applying the right tool at the right time and in a holistic manner, remarkable benefits can be recognized.

GROUP TECHNOLOGY

Group Technology is one of the fundamental tools of any organization moving away from functional (departmentalization) manufacturing and into a cellular-manufacturing environment. Rather than having all products flow through the same pipeline, Group Technology identifies the commonality between product types or process requirements, and then groups them into logical work cells.

Figure 3-2 shows some symbols used for simulating parts and processes. First, imagine these parts travel through the shop based on whatever machine is open and capable of processing them. In any given week, the flow through the shop might look like Figure 3-3. A viable alternative to this approach is grouping the materials together that tend to flow the same way or require similar equipment. The result is shown in Figure 3-4. Even though shared

Table 3-1. Toyota Production System (TPS) summary

Relationship to Six Sigma	Like Six Sigma, TPS focuses on productivity, delivery, and cost issues, as well as opportunities to improve quality.
Who needs and uses it	The vice president of operations, production managers, and value-stream leaders use it. TPS is particularly useful in assembly operations.
Cost	Moderate for a model line/high for an entire facility
Strengths	TPS reduces lead times and inventory costs and improves productivity and space utilization.
Limitations	TPS is limited where raw-material lead times, product, or order cycles are very hard to predict.
Process complexity	Moderate to difficult
Implementation time	6 months (for most model lines) 5 years (for most job shops)
Additional resources	See Bibliography
Internet search key words	TPS, Toyota Production System
Internet URL	www.toyotaproductionsystem.net

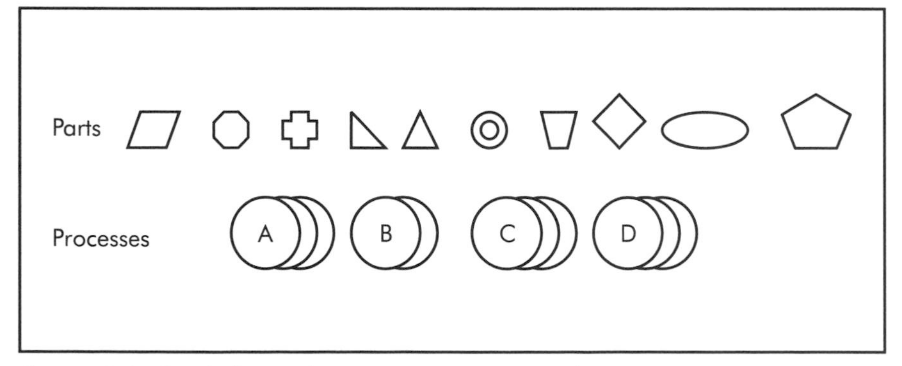

Figure 3-2. Symbols used to represent parts and processes.

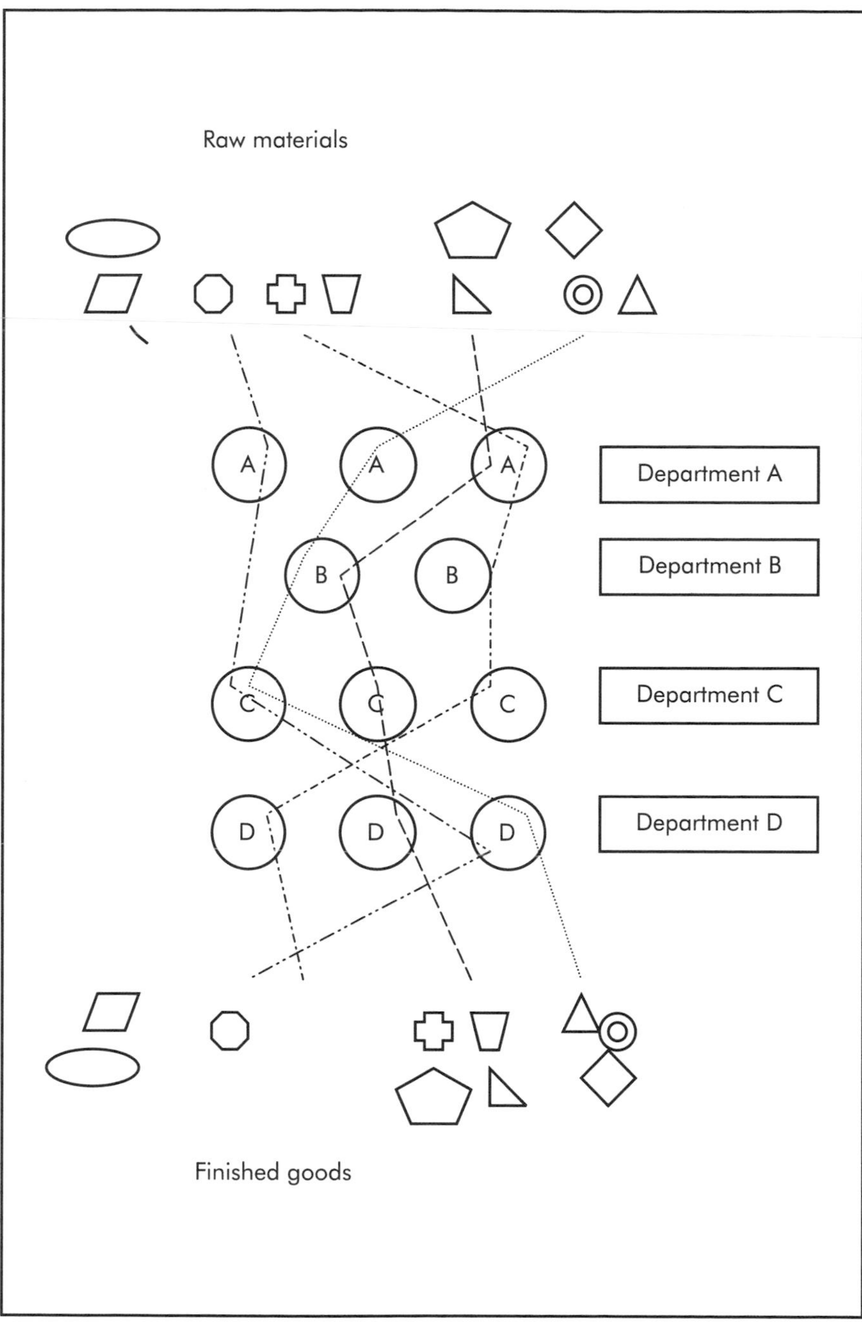

Figure 3-3. Group Technology examines typical manufacturing flow.

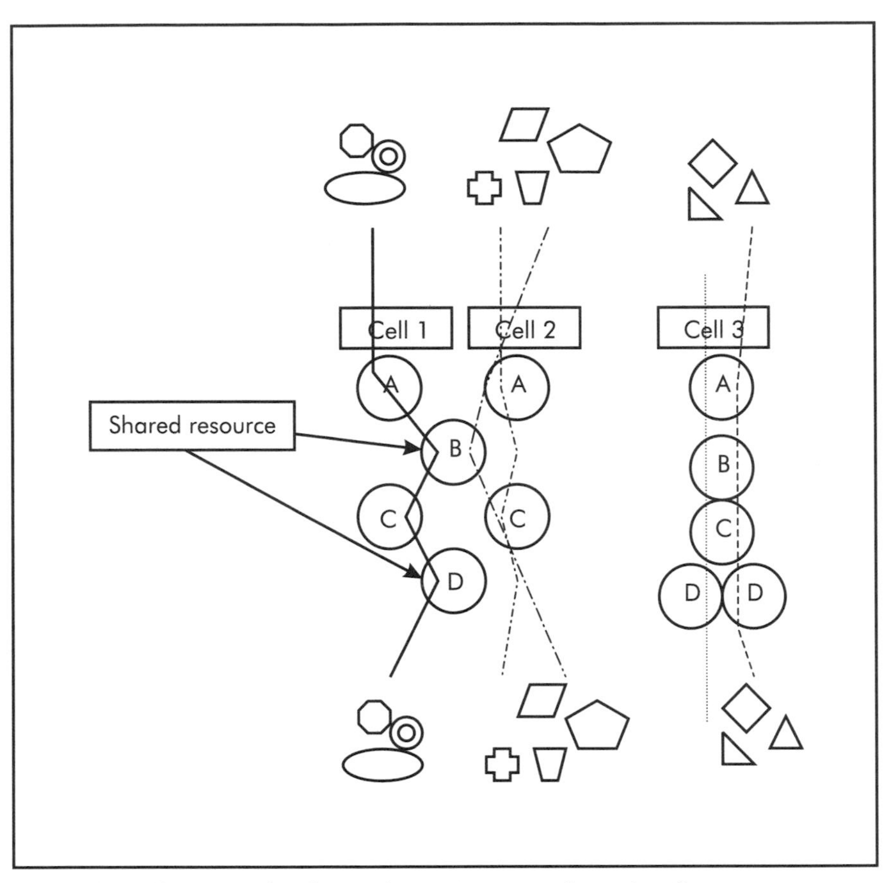

Figure 3-4. Group Technology alternative manufacturing flow.

resources still have to be dealt with, product velocity is increased, and items that tend to languish around in a functional layout, plugging up the flow, can now be segregated. Otherwise, it would be much like putting a Model T on the Autobahn. It just slows everybody down. This is not saying anything against that "Model T odd job," which takes extra time. You just don't want it constraining the rest of the flow if there is a way to process it outside the mainstream. Taking all those Model Ts off the Autobahn and putting them together on a little side road would make it safer for everybody. That is the goal of Group Technology, and it is summarized in Table 3-2.

Table 3-2. Group Technology summary

Relationship to Six Sigma	By dividing parts into logical value streams, group technology fosters greater control.
Who needs and uses it	Production managers
Cost	Low if done as virtual cells/moderate if facility is reconfigured
Strengths	Group technology allows better flow, and reduction of paperwork and nonvalue-added activities.
Limitations	Reduced flexibility in some cases
Process complexity	Easy to moderate
Implementation time	1–6 months
Additional resources	See Bibliography
Internet search key words	Cellularization, group technology
Internet URL	www-iwse.eng.ohio-state.edu/ISEFaculty/irani/research/overview_of_prof_iranis_research.htm

TOTAL QUALITY MANAGEMENT

Whether called Total Quality Control (TQC) or Total Quality Management (TQM), the tool was one of the foundational elements of the Japanese turnaround that began in the 1960s. This set of techniques has been spreading throughout America for over two decades now. Dozens of books and hundreds of trade-journal articles have been written about it. Quality-circle teams were formed and taught to use the Ishikawa diagram for problem solving. The teams learned to use the Pareto chart to identify the greatest opportunity for improvement. They were trained in how to plot and interpret histograms, X-bar and R charts, and control charts in hopes they would be better able to understand and control processes or brainstorm potential solutions.

If Lean Manufacturing is a mature stage of the Toyota Production System, then Six Sigma is the evolutionary result of the work initiated by Deming, Juran, and others in Japan in the early 1950s: the TQC/TQM movement that took flight in Japan's early transformation.

Some of the many and varied tools of TQM are discussed later, but an example of one TQM tool that often gets overlooked is the scatter diagram. Use of this tool might be helpful in the imaginary concrete-block manufacturing plant example. Trying to control a recipe like concrete slurry is an excellent application for a scatter diagram. For example, after measuring what happens as water is added or subtracted from the slurry, the ideal amount can be found by plotting the results on a scatter diagram. From the simple (fictitious) results shown in Table 3-3, it appears that adding 7 oz (198 g) of water to 1 lb (0.5 kg) of sand is the ideal amount. The critical step is to control all other variables during a test like this. Otherwise it cannot be assured that what is being tested is the only variation. For example, if mixing times are adjusted at the same time as performing a test on water content, the results will be suspect.

Table 3-3. Scatter diagram showing the effect of adding or subtracting water from the slurry

Strength after drying 1,000 psi (Pa)										
10 (6.9)										
9 (6.2)										
8 (5.5)										
7 (4.8)										
6 (4.1)										
5 (3.5)							(X)			
4 (2.8)						X		X		
3 (2.1)				X	X				X	
2 (1.4)			X							
1 (0.7)		X								X
Ounces of water per pound of sand (g of water per kg of sand)	1 (63)	2 (125)	3 (188)	4 (250)	5 (313)	6 (375)	7 (438)	8 (500)	9 (563)	10 (625)

Scatter diagrams also can be used as a predictive tool. For example, to minimize fuel cost for a cross-country drive, a scatter diagram such as the one in Table 3-4 could be set up to find the optimum travel speed to maximize fuel economy and better predict gas consumption.

Table 3-4. Scatter diagram comparing speed relationship to gas consumption

Miles/gal (km/L)										
20 (8.5)										
19 (8.0)										
18 (7.7)	X									
17 (7.2)		X								
16 (6.8)			X							
15 (6.4)				X						
14 (6.0)					X					
13 (5.5)						X				
12 (5.1)							X			
11 (4.7)										
Average travel speed, mph (km/h)	35 (56)	40 (64)	45 (72)	50 (80)	55 (88)	60 (97)	65 (105)	70 (113)	75 (121)	80 (129)

Next, to show the relationships of two variables, speed and fuel type, the results can be overlayed as shown in Table 3-5 to see the benefits of premium fuel.

There are literally hundreds of tools, such as histograms, scatter diagrams, and sigma calculators, and the challenge is always to find the tool that most simply illustrates where you are now in relationship to the goal or ideal condition you are seeking to achieve. A summary of the Total Quality Management tool is shown in Table 3-6.

Table 3-5. Scatter diagram comparing speed and fuel type to gas consumption

Miles/gal (km/L)										
20 (8.5)										
19 (8.0)										
18 (7.7)	RP									
17 (7.2)		RP								
16 (6.8)			R	P						
15 (6.4)				R	P	P				
14 (6.0)					R	R	P			
13 (5.5)							R			
12 (5.1)										
11 (4.7)										
Average travel speed, mph (km/h)	35 (56)	40 (64)	45 (72)	50 (80)	55 (88)	60 (97)	65 (105)	70 (113)	75 (121)	80 (129)

R = regular octane P = premium octane

Table 3-6. Total Quality Management (TQM) summary

Relationship to Six Sigma	The tools used in TQM and Six Sigma programs are virtually the same.
Who needs and uses it	Project coordinators, black belts, green belts, and project teams
Cost	Low (depending on the number of tools used)
Strengths	Helps identify opportunities for improvement and problem-solving solutions
Limitations	Requires training to learn how to use it.
Process complexity	Depends on the tools selected
Implementation time	1 week–3 years, depending on the company size and number of tools used
Additional resources	See Bibliography
Internet search key words	Total Quality Management, Hoshin, TQM
Internet URL	www.apqc.org

THEORY OF CONSTRAINTS

The Theory of Constraints, developed and taught by Dr. Eliyahu Goldratt and the Abraham Y. Goldratt Institute, is based on identifying and removing bottlenecks. It presents a critical-thinking approach to solving problems of any magnitude or description and is a powerful mechanism for positive change (improvement) and eliminating organizational constraints. Its five-step process is similar to DMAIC:

1. Identify the system's constraint.
2. Decide how to exploit the system's constraint.
3. Subordinate everything else to the above decision.
4. Elevate the system's constraint.
5. If in the previous steps a constraint has been broken, go back to the first step, but do not allow inertia to cause a system constraint.

In other words, find out where the bottleneck is and remove the reasons for it. For example, identifying where the flow of money is constricted within an organization requires a difficult self-examination and departure from the natural tendency to self-protect. Here, evident "suboptimization" is made visible by critical-thinking techniques. Results are measured, and inefficiencies are often a manifestation of what has been measured for decades, particularly in smaller organizations. Having managers expend energies to make the weekly, monthly, or quarterly departmental target numbers can end up costing the company the greater good, an example of suboptimization.

The process of problem identification and critical thinking is important to improvement. By asking the non-obvious questions, teams can often discover easier answers and clearer paths to a solution. An obvious solution is ordinarily overlooked if it is assumed all current conditions are a given or are unchangeable.

For example, in the case of the fictitious concrete-block company, there is a problem related to unacceptable variation in weight. One of the prerequisites to making blocks of uniform weight is to have proper percentages of water, sand, and concrete. An input that could negatively affect the outcome could be that the machine operator does not take time to calculate the exact percentage of ingredients to add, or in efforts to speed up the mixing cycle, is undermixing the material.

The approach used in the Theory of Constraints calls for development of a clear statement of the current reality by using a problem-analysis tool called the current-reality tree (CRT) (Figure 3-5). Similar to a flow chart, it makes visual what may or may not be clear about the process. Then, like Value Stream Mapping, which will be discussed later, a future-reality tree (FRT) (Figure 3-6) is developed for the cement-block process to define a favorable alternative to the unacceptable current condition. Many times conflicts are hidden until the effort is made to graphically represent them. In this case, the conflict between the required speed of the operator and his or her need to calculate the proper recipe or mix-

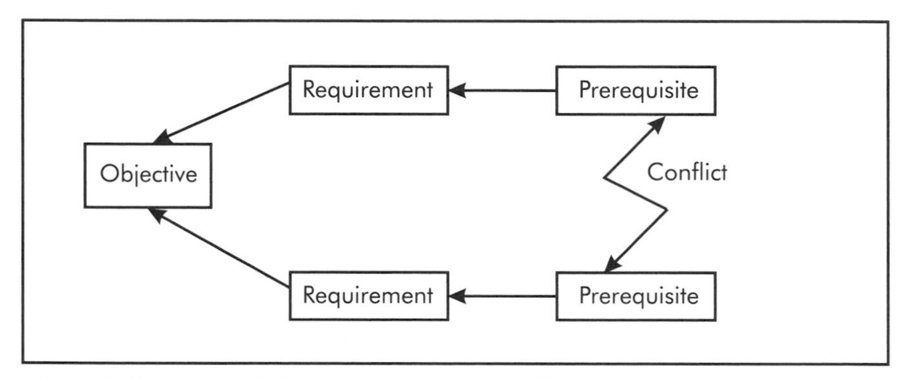

Figure 3-5. Theory of Constraints current-reality tree.

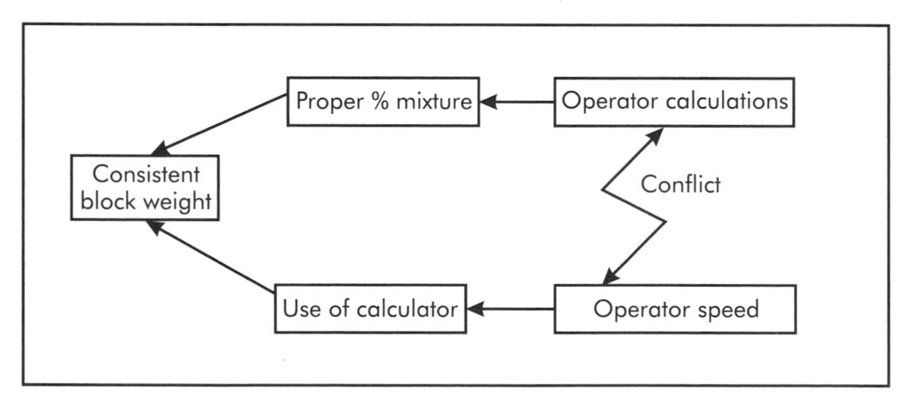

Figure 3-6. Theory of Constraints future-reality tree.

ing time could impact the resultant quality of the concrete blocks produced downstream.

By asking the tough and less obvious questions, sometimes a simple solution can be discovered or created. For example: in the cement block case, is a calculator the only tool for determining the proper concrete recipe mixing? Would a quick-reference wall chart do as well? Could the dry mixture be pre-measured to eliminate part of the calculation? Could the process be automated? Could a computer take the human element out of the process, performing the calculations for the operator? The point is, the most important question is not always asked, "Is this the only way to do this?"

The Theory of Constraints dovetails well with Six Sigma because as you try to create changes that will result in long-lasting gains, there is usually a simple answer that will result in changes that are easier to implement and sustain.

Drum-buffer-rope (DBR) is another Theory of Constraints tool. It is a methodology and subset of production-planning techniques used to help minimize the negative effect of a constraint or bottleneck somewhere within the organization. Drum-buffer-rope has proven to be the best method to manage physical processes. DBR attempts to find solutions that offer maximum throughput, shortest flow times, better predictability, and minimum inventory levels.

DBR focuses attention on what needs to be managed and allows the rest of the system to subordinate (pace) itself to the constraint. In other words, why produce on an upstream machine if the downstream constraint cannot accept it anyway? This is a hard concept for many managers to accept, knowing what a particular piece of equipment cost the company. It is hard for a manager to walk out on the shop floor and not hear that expensive equipment pounding out parts (or whatever sights or sounds reflect optimum production).

People think of Toyota as having no inventory. Yet it does have inventory. Inventory is simply used in a way that is counter-intuitive. If its workers have a problem or constraint that cannot be solved, then Toyota must use inventory like any other company to insulate against missed delivery dates until there is a solution to the problem or the constraint is eliminated.

The difference in the way Toyota deals with inventory is a matter of where it locates or stores the inventory. It is not hidden in a rack somewhere. If at all possible, inventory is put right in the middle of the floor, in everyone's way. This encourages people to think about the problem because they must navigate around that problem every day until it is solved.

A job shop might similarly use DBR where there are hard-to-predict sales fluctuations that could cripple the company's ability to react overnight to a spike or drop in sales. Like a fish tugging on the end of a fishing pole, DBR is immediate and very easy to interpret (Figure 3-7). The *drum* is a factor that controls the pace and synchronization of the value stream. Sometimes the point of sale functions as the drum. The *buffer* consists of either outsourced material or in-house (work-in-process) materials. Each of these may be somewhere in transit or in the manufacturing pipeline.

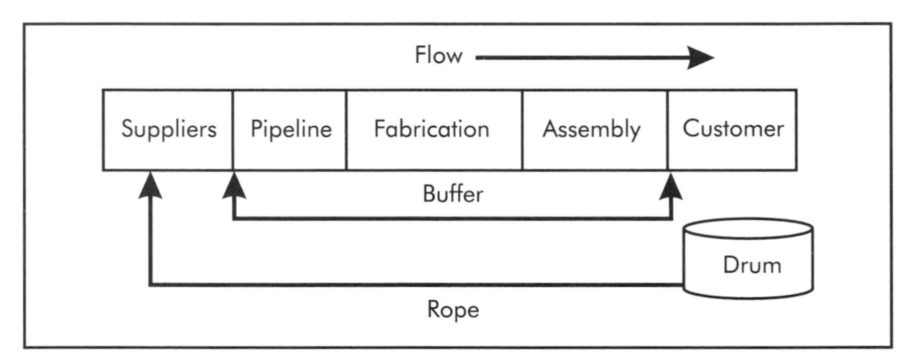

Figure 3-7. Drum-buffer-rope (DBR) conceptual graphic.

The purpose of the buffer can be to deal with the uncertainty of customer demand or to accommodate problems or bottlenecked resources that cannot be immediately resolved. The *rope* controls the release of additional product into the system. In larger organizations, this will often be managed by the Material Requirements Planning/Equipment Requirements Planning (MRP/ERP) supply-chain management system. It also can be managed through the use of Kanban cards (signals to make or move something).

The general opinion is that DBR has a good deal more application within original-equipment manufacturers (OEMs) than in job-shop operations because DBR assumes you have a good idea of what you are going to make or sell. So far, its application within the make-to-order environment has been limited. However, many job shops are beginning to recognize the value of "leaning out" inventory buffers. Regardless of what jobs are in the backlog, the shipping department *pulls the rope* signaling the first process to introduce another job into the value stream. Everything runs at the pace of the constraint, and all energies are focused on removing that constraint. The Theory of Constraints is summarized in Table 3-7.

TOTAL PRODUCTIVE MAINTENANCE

Trying to become World Class while operating with second-class machines is an impossible dream. This does not mean that

Table 3-7. Theory of Constraints (TOC) summary

Relationship to Six Sigma	Operating at a Six-Sigma level of performance requires considering commitments and how closely they can be fulfilled. TOC assists in defining the company's true capabilities and capacities to avoid overcommitting and under-delivering.
Who needs and uses it	Project coordinators, black belts, green belts, project teams
Cost	Low (depends on the number of tools used)
Strengths	Helps identify opportunities for improvement and problem-solving solutions
Limitations	More challenging in make-to-order shops
Process complexity	Moderate
Implementation time	1 month–3 years, depending on project complexity and the number of tools used
Additional resources	See Bibliography
Internet search key words	TOC, Theory of Constraints, Goldratt
Internet URLs	www.eligoldratt.com/ www.pdinstitute.com/home.html

a company needs to go out and buy new machines. Total Productive (or predictive) Maintenance (TPM) is an effective, time-proven system. In some cases, it has shown such remarkable results as reducing machine breakdowns to nearly zero while increasing worker productivity by as much as 150%.

Since first being introduced in Japan by Seiichi Nakajima, TPM has played a major role in a worldwide revolution in plant maintenance. In 1950, Nakajima, then vice chairman of the Japan Institute of Plant Maintenance, began studying preventive maintenance as practiced in the United States. In 1971, he developed TPM and introduced it into Japan.

TPM is a set of equipment-management activities. The goal of TPM is total prevention of quality defects and machine breakdowns. Additional motivating factors include reducing set-up time and material waste, along with eliminating equipment adjustments and increasing operator safety.

When maintenance activities are the sole responsibility of the maintenance department, companies are seldom successful in achieving zero equipment breakdowns. TPM requires company-wide participation. The persons most likely to notice equipment problems or symptoms are not the maintenance team, but the operators working with the machines on a daily basis. A qualified operator should listen to his or her machine just as a racecar driver listens to the engine and understands what is going on deep inside it.

TPM requires total cooperation between production and maintenance personnel and obviously requires additional training for both. The three main characteristics of a sound TPM program are:

- maintenance of normal conditions,
- early discovery of abnormalities before a breakdown occurs, and
- prompt response to any condition needing attention.

Total Productive Maintenance:

- maximizes overall equipment effectiveness,
- establishes a thorough system of preventive maintenance over the entire life expectancy of the equipment,
- involves every employee from top management to production people,

- is promoted through "motivative" rather than "punitive" management, and
- is implemented by engineering, operations, and maintenance departments.

It is essential that upper management be fully committed to TPM and understand that a paradigm shift is required for it to be effective. Management must be the driving force, and value-stream managers must ensure that everyone is viewed as part of the TPM team to achieve improvement in total productivity.

The machine operator will be the person(s) most affected by the changes implemented in a TPM program, especially in the early stages of transformation. Downtime causes must be recorded. Keeping track of part failures and reasons for downtime is another responsibility not previously held by the operator. Most operators already have significant record-keeping responsibilities and understandably may resist having to add more. Although this is clearly a nonvalue-added activity, accurate data collection is necessary if predictive tools are to be used to eliminate downtime.

Operators will also need to learn new skills related to cleaning and inspection of the equipment. This may include limited noninvasive maintenance, part replacement, lubrication, and adjustments. Operators may even work closely with the maintenance technicians when major repairs are performed. By participating in this way, operators begin to develop basic diagnostic skills so they can offer responses that are more meaningful in the future.

Implementing change on a smaller scale can help in first developing a positive model and testing the skills required, rather than trying to implement a massive change plant-wide. Here are some critical steps:

- Clean and inspect: All machine areas are cleaned thoroughly and visually inspected to make sure minor defects are recorded for needed repairs. All lubrication points are identified. Teams generate ideas to facilitate easier cleaning and visual inspection. Contamination opportunities (that is, oil drips) are eliminated or repaired.
- Observe machine condition: Every part of the machine is put on a schedule of inspection. Any questionable component that

could potentially cause a defect is repaired immediately. Minor abnormalities and deterioration are restored to proper or original condition.

- Develop preventive maintenance checklists: Like the owner's manual that comes with a new car, maintenance checklists are often provided by machine manufacturers. If these checklists and manuals are vague, too generic, or long forgotten, the TPM team should develop their own. Lubrication types and amounts should be clearly stated within the documents.
- Develop a preventive maintenance schedule: Once the checklists have been created, a scheduling system for preventive maintenance should be developed. New software packages are available to take some of the more mundane scheduling tasks away from people, giving it to the computer instead. Try to make things visual rather than depending on a people-driven system.

A very simple, yet brilliant example of a preventive-maintenance (and 5-S which will be discussed later) scheduling system is being used at a company in Southern California. The team came up with a mail-slot-like cluster of little trays, each having two columns, each with 31 slots representing the days of the month. Each slot contained a card directing the operator (and any roving maintenance personnel) to perform key checks and lubrication or cleaning activities for that day.

Once the tasks were performed, the card was moved to the other column (see Figure 3-8). When all the cards had been moved on the last day of the month, the team started again at the top and began moving them back the other way again. Value-stream leaders (or the maintenance manager) could make their way through the shop, and with a glance determine whether the preventive-maintenance procedures had been performed for the day. It is a visual system that is both easy to maintain and easily trained.

An important goal of any TPM program is to make it easy to monitor the machine condition. If this is hard to do, it won't get done. Some examples of improvements a company can make are relocating hard-to-reach grease fittings or mounting lubrication points in more accessible locations, co-locating grease fittings, and replacing metal belt guards with clear-plastic versions so that belts can be more easily inspected for wear.

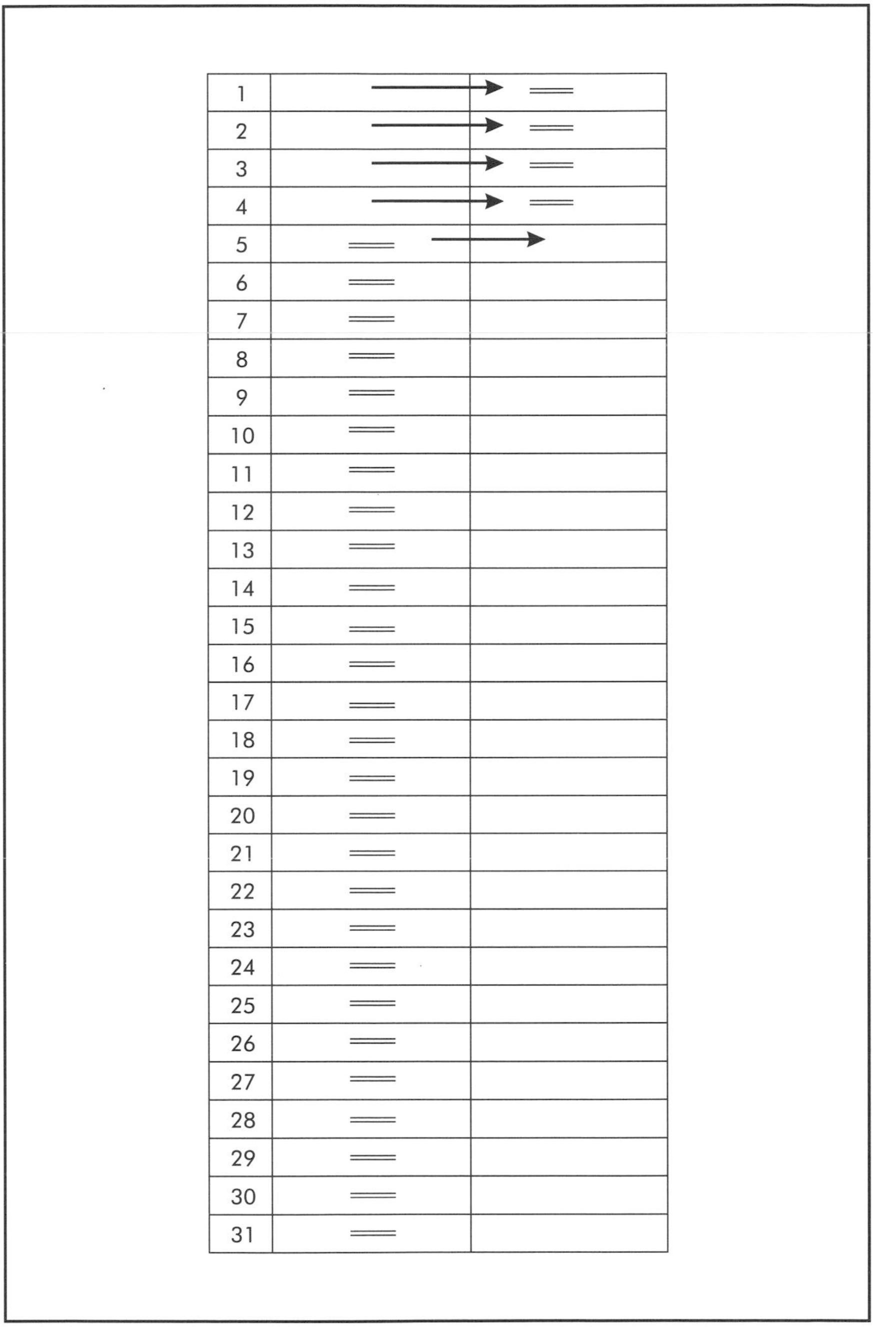

1		
2		
3		
4		
5		
6		
7		
8		
9		
10		
11		
12		
13		
14		
15		
16		
17		
18		
19		
20		
21		
22		
23		
24		
25		
26		
27		
28		
29		
30		
31		

Figure 3-8. Visual system for preventive maintenance.

Some typical benefits realized by companies implementing a comprehensive TPM program include:

- productivity up 140%;
- value-added per person increased 117%;
- rate of operation increased by 17%;
- breakdown rate decreased by 98% per year;
- quality claims reduced by an average of 42%;
- cost reductions of 30% with no decrease in productivity;
- reductions in maintenance costs averaging 25%;
- energy savings of 30%;
- finished goods inventory reduced by 50%;
- safety incident improvements; and
- morale improved by as much as 215%.

The dollars lost due to downtime, set-up time, and quality problems all relate to poorly maintained equipment. Having a meaningful and comprehensive TPM program can go a long way toward insuring machine capability and reliability. Table 3-8 summarizes TPM.

BALDRIGE QUALITY PROGRAM

The Malcolm Baldrige National Quality Award (MBNQA) was established by the Department of Commerce, National Institute of Standards and Technology (NIST) in 1987 to promote Total Quality Management as an increasingly important approach for improving the competitiveness of American companies. The MBNQA has evolved from a quality focus to an award for performance excellence. Like a meaningful Six-Sigma approach, the holistic nature of the award criteria measures not only product quality, but also leadership strategy and communication of that strategy. It includes not only whether the customer gets what they want, but whether the long-term viability of the provider of that product is ensured through profitability and the proper management of the human resources so critical to any business.

In the late 1980s, President Reagan's cabinet worked to establish a new national effort to improve quality. The award is named after Commerce Secretary, Malcolm Baldrige, who was killed in a rodeo accident. The guidelines for winning this award include a

Table 3-8. Total Productive Maintenance (TPM) summary

Relationship to Six Sigma	Machine reliabilty, dependability, and repeatability are key to any Six-Sigma effort.
Who needs and uses it	Project coordinators, black belts, green belts, project teams
Cost	Moderate, depending on current condition of machines
Strengths	Any company can initiate these tools.
Limitations	Implementation by itself will only slightly improve company performance.
Process complexity	Moderate
Implementation time	1 month–3 years, depending on current machine conditions
Additional resources	See Bibliography
Internet search key words	TPM, machine reliability, maintenance programs
Internet URLs	www.reliabilityweb.com/index.htm www.plant-maintenance.com/index.shtml www.pmcrae.freeserve.co.uk/oeecalc.html

requirement for evidence that the company has involved everyone at every level in the quality effort. Xerox Corporation was among the first few companies to win the award.

The MBNQA's Criteria for Performance Excellence is scored on a scale of 1,000 total points. The categories are refined periodically. The distribution of Year 2000 point values is shown in Table 3-9.

The actual criteria cover 16 pages of text and the intent and approach are sound. Any company seeking to examine itself from a quality standpoint would do well to use it as a Baldrige auditor might.

The process of conducting a first-time audit is commonly referred to as a gap audit, meaning it reveals the gaps in performance. Having an outsider perform a gap audit can understandably be a

Table 3-9. Abbreviated Malcolm Baldrige criteria

Category Number	Items	Point Value Item	Point Value Total
1	Leadership		125
	Organizational leadership	85	
	Public responsibility and citizenship	40	
2	Strategic Planning		85
	Strategy development	40	
	Strategy deployment	45	
3	Customer and Market Focus		85
	Customer and market knowledge	40	
	Customer satisfaction and relationships	45	
4	Information and Analysis		85
	Measurement of organizational performance	40	
	Analysis of organizational performance	45	
5	Human Resource Focus		85
	Work systems	35	
	Employee education, training, and development	25	
	Employee well-being and satisfaction	25	
6	Process Management		85
	Product and service processes	55	
	Support processes	15	
	Supplier and partnering processes	15	
7	Business Results		450
	Customer focused results	115	
	Financial and market results	115	
	Human resource results	80	
	Supplier and partner results	25	
	Organizational effectiveness results	115	

stressful process, much like having guests go through your closets. Yet the opportunity for problem identification (one of the first steps in the Six Sigma and Kaizen approaches) can help your company quickly begin working on the right things. The Baldrige quality program is summarized in Table 3-10.

Table 3-10. Malcolm Baldrige quality program summary

Relationship to Six Sigma	Many of the same principles as Six Sigma apply. It could compete for resources if initiated in parallel with Six Sigma.
Who needs and uses it	Director of quality, quality assurance managers, value-stream leaders
Cost	High
Strengths	Provides structure
Limitations	It could give a false sense of security that your company has arrived when there is really no finish line.
Process complexity	Moderate to difficult
Implementation time	1–3 years
Additional resources	See Bibliography
Internet search key words	Baldrige, quality systems, NIST
Internet URL	www.quality.nist.gov

DEMING PRIZE

The Deming Prize was introduced in 1951 by the Japanese Union of Scientists and Engineers (JUSE) to honor Dr. W. Edwards Deming and promote the implementation of quality control in Japan as it struggled to recover from the war.

Like the Baldrige criteria, seekers of the Deming Prize focus on a holistic approach to quality. Independently verified by experienced auditors, the quality system of each company seeking certification is scrutinized to ensure that what has been documented has depth and can be sustained. The following is a summary of the critical-to-quality list as compiled by the Deming Application Prize Subcommittee:

1. Policies: Includes quality and quality-control policies and their clarity, methods, and processes for establishing policies; relationship of policies to short- and long-term plans; communication (deployment) of policies; and executive and manager leadership.

2. Organization: Includes appropriateness of the organizational structure for quality control, employee involvement, clarity of authority and responsibility, and relationships with associated companies (group companies, vendors, contractors, etc.).
3. Information: Includes the appropriateness of collecting and communicating external and internal information, status of applying statistical techniques to data analysis, how information is retained and used, and the use of computers for data processing.
4. Standardization: Includes appropriateness of the system of standards; procedures for establishing, revising, and abolishing standards; content of standards; status of using and adhering to standards; and status of systematically developing, accumulating, handing down, and using new technologies.
5. Human-resource development: Includes the constant reporting and evaluation of the status of education and training plans and results, quality consciousness, supporting motivational programs for self-development and self-realization, use of statistical concepts and methods, quality-control-circle development, and supporting development of human resources in associated companies.
6. Quality-assurance activities: Includes monitoring how well the quality-assurance system is being managed and the status of quality-control diagnosis; new product and technology development; process control, analysis, and improvement (including process capability studies); inspection; quality evaluations and audits; the managing of production equipment, measuring instruments, and vendors; product usage, disposal, recovery, and recycling; and the measuring of customer satisfaction, product reliability, safety, product liability, and environmental protection.
7. Maintenance/control activities: Includes the rotation of management (plan, do, check, act) cycle, methods for determining control items and levels, in-control situations (status of control-chart use and other tools), and the status of operating management systems for cost, quantity, delivery, etc.

8. Improvement activities: Includes methods of selecting themes (important problems and priority issues), linkage of analytical methods and intrinsic technology, use of statistical methods for analysis, analysis of results, and status of confirming improvement and transfer of that improvement to maintenance/control activities.
9. Effects: Includes measuring and controlling intangible and tangible effects (such as quality, delivery, cost, profit, safety, and environment), methods for measuring and grasping effects, customer and employee satisfaction, and influence on associated companies and local and international communities.
10. Future plans: Includes the status of grasping current situations, future plans for improving problems, projection of changes in social environment and customer requirements, management vision and long-term plans, continuity of quality-control activities, and concreteness of future plans.

Deming's list of 14 points (closely related to his list of deadly diseases) should be used by the management team to look into the mirror and recognize areas where they need to modify their behavior. Without this, there is little chance that the company's culture is ready for sustainable change.

Deming stated that 85% of all performance problems are rooted in the system. Who is in charge of the system? Clearly it is management. Deming's list is as follows:

1. Create consistency of purpose toward improvement of product and service.
2. Adopt a new philosophy (because we are in a new economic age).
3. Cease dependency on inspection; build quality in.
4. Don't focus on the price tag; minimize total cost.
5. Build long-term relationships built on loyalty and trust.
6. Improve the system constantly to decrease costs.
7. Institute training on the job.
8. Institute leadership and focus on helping people do a better job.
9. Drive out fear; break down barriers between departments.
10. Eliminate slogans, quotas, and management by numbers.

11. Eliminate barriers that rob hourly workers of pride in workmanship.
12. Abolish the annual merit rating.
13. Institute a vigorous program of education and self-improvement.
14. Put everybody in the company to work to accomplish the transformation.

Like any great sports coach or personal trainer, Deming had his critics, but the results speak for themselves. His approach has not been embraced by everyone, but as his most famous quotation clearly states: "You do not have to do this. Survival is not mandatory." Deming gave the customer what they needed to hear, not always what they wanted to hear. The Deming Prize approach is summarized in Table 3-11.

ISO 9000 QUALITY SYSTEM MANAGEMENT

ISO 9000 has gotten a bad reputation of late. In the late 1980s and early 1990s, ISO 9000 compliance was courted by many company managers. Everyone seemed enamored with it. Thousands of consultants sprang up overnight to help companies get certified.

Then, after certification, many companies realized that in their haste, they had over-documented processes, to the degree that they now had a hard time making improvements without violating a policy, procedure, or work instruction. The cost of improvement became prohibitive because people felt painted into the corner by this documentation. Companies found themselves with a full staff of document-control people, but no one was actually spending time looking for improvement opportunities. Instead, everyone was spending their time trying to keep their respective "ducks in a row" in expectation of the next surveillance audit.

In the most recent revision of the standard, the scope of ISO 9001:2000 has expanded to include issues such as applicable regulatory requirements, customer satisfaction, continual improvement, and permissible exclusions. This should help reduce the heartburn related to overly constrictive procedures and policies.

This is not an indictment of the ISO 9000 system—nothing could be further from the truth. The structure of the ISO program is

Table 3-11. Deming Prize summary

Relationship to Six Sigma	Many of the same principles apply to the Deming Prize. The two programs could compete for resources if initiated in parallel.
Who needs and uses it	Director of quality, quality assurance managers, value-stream leaders
Cost	High
Strengths	Provides structure
Limitations	A few companies won this award and later struggled or failed soon afterward, raising questions about the sustainable impact of the approach.
Process complexity	Difficult
Implementation time	1–3 years
Additional resources	See Bibliography
Internet search key words	Deming, Deming Prize
Internet URLs	www.deming.org/index.html caes.mit.edu/deming/tdv.html www.amanet.org www.qualitydigest.com/html/resource.html

very sound, but ISO consultants are too often rewarded only when they get the job done quickly. In an effort to make registration a fast and a sure thing, a consultant uses what he or she knows works, and this usually means applying a time-tested, boilerplate template to every company. Long after the consultant has gone, the registrar will be coming in to perform audits on a program that nine times out of 10 meets the registrars needs rather than those of the company being audited.

ISO 9000 does not prohibit a company from focusing first on customer needs. However, the approach taken during the preparation for the registration audit is rarely about meeting customer needs. It is about passing a test that requires describing what the company does (whether good or bad), and then proving it can be done (well or poorly) consistently. In the early days of ISO 9000, a company could describe what it did and probably get certified, even if its product was a concrete life preserver. As long as the company bought the concrete from an approved vendor and molded it according to a clearly defined procedure, it made no difference whether or not the end user was satisfied with the result. If the company made it wrong, ISO 9000 would help ensure that it was made wrong every time. Thankfully, the latest refinements to the ISO standard have addressed some of these important issues. Reduced from 20 paragraphs (or articles) down to eight, the standard takes a more holistic look at the entire business, not just the production of parts.

Similar to both the Baldrige and Deming approaches, the auditor and audit verification process for ISO 9000 is overseen by an international body known as the Registration Accreditation Board (RAB). The International Organization for Standardization (ISO) has worked to standardize many things, including identification of what constitutes a sound quality system.

The ISO 9000 criteria in the 2000 version of the standard do not necessarily dictate how a company should be run (as many people believe). The criteria simply seek to have a company clearly document how it operates, and how it addresses over 100 critical business principles. These requirements define the areas in which documentation or evidence related to policy-procedure-work instructions or records will be examined during an independent third-party audit.

The 2000 revision strengthens the requirements for the control of knowledge and design. As the Information Age matures, it is reasonable from a quality perspective that control of knowledge in all its forms will be strictly managed. Tribal knowledge no longer works in this age of a highly mobile work force.

Appendix A shows an interpretation of the 2000 version of the ISO 9000 standard. The 1994 version, which included a distinction between 9001, 9002, 9003 is no longer valid, and companies have until the end of 2003 to update their systems to the new standard if they wish to maintain their certificates.

Regardless of these latest improvements to the standard, ISO-certified companies will have no greater chance of positive impact than they did with the old version if the structure continues to be applied as a one-size-fits-all template. Each company must recognize its unique needs and not race for registration. Rather, a company should seek a program that provides for realistic controls while allowing for the maturation and improvement process. If not, the confines of a rigid program will impede rather than foster growth and incremental gains. ISO 9000 is summarized in Table 3-12.

Table 3-12. ISO 9000 summary

Relationship to Six Sigma	ISO 9000 can support the Six Sigma process well if viewed as a means to document improvement processes.
Who needs and uses it	Executive management, director of quality, value-stream managers
Cost	High
Strengths	Provides structure
Limitations	The ISO structure does not contain or teach tools for how to improve.
Process complexity	Moderate to difficult
Implementation time	1–2 years
Additional resources	See Bibliography
Internet search key words	ISO 9000, quality systems
Internet URLs	www.iso.ch/iso/en/ISOOnline.frontpage www.isoeasy.org

4

Step One: Define

Understanding where to start is possibly the most difficult step, and this chapter is designed to help sort out that question. Approaches will range from use of the Kano model to find out what excites a customer to buy, to the use of quality function deployment (a fancy way of saying "get it right the first time"). By examining a company's operation and using a graphic representation like the suppliers, inputs, processes, outputs, and customers (SIPOC) process, the Six-Sigma approach can help visualize the inputs and outputs of each process and assist in determining where unwanted variables are introduced. Discussing how process yield can be measured will help define opportunity, and therefore future projects.

Listening to the customer is addressed in this chapter. An example of a customer survey is offered to use in gaining customer feedback. An affinity diagram is used to examine the interrelationship of activities, things, or people.

A short discussion of Hoshin Kanri (executive decision making) follows a review of a popular 1980s technique called "management by walking around." Does this technique still have value? The answer is yes!

Especially important to job shops and make-to-order shops, the discussion on product, quality, routing, support systems, and time (PQRST) will assist in identifying discrete value streams within an organization. Finally, the development of a clear and concise team charter (project mission statement) will demonstrate the value of defined goals and objectives.

KANO MODEL

What satisfies a customer? Or looking at it the other way, what do you want as a customer? What do you need? What excites you to buy? These questions sum up the idea behind the Kano model developed by Nortitaki Kano. His theory is that there are four emotional quadrants (Figure 4-1) into which customers can be categorized.

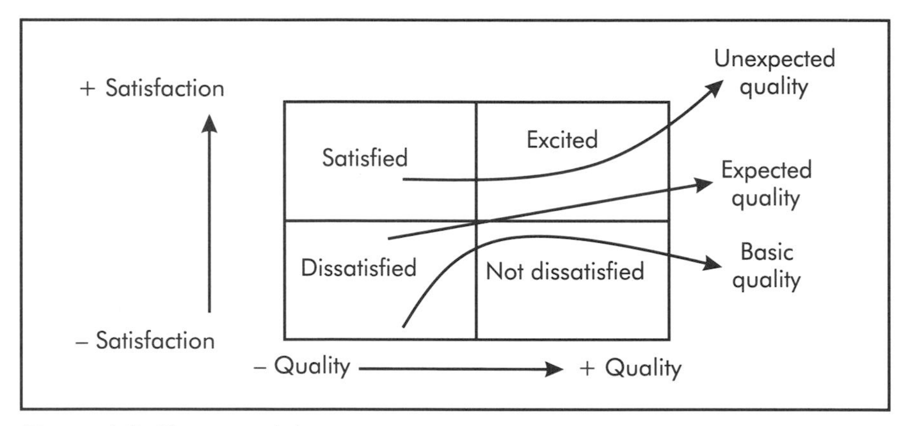

Figure 4-1. Kano model.

What does it take to excite a customer? How about zero defects? No. A customer expects zero defects. If a customer buys a new gas barbeque and finds no defects when it is put together, he or she will not be dissatisfied. Yet, since everything is expected to work, neither is he or she necessarily satisfied. However, if the barbecue is purchased and the store manager offers to assemble it for free, this is very satisfying. Then, if he throws in a coupon for a free tank of fuel, the customer will be excited.

Where basic quality is simply met but not exceeded, a customer will probably be neither dissatisfied nor particularly satisfied, and certainly not excited about quality. Exciting quality features can often be simple things that are simply unexpected, like an insulated wrap for your morning latte.

Innovation is what drives excitement. Innovation comes from competitive pressure. Customers will never tell a company what they really want. They may not even know that they want it, but the minute a competitor gives it to them, then they can't live without it.

It would be easy to say that Six Sigma only applies to original-equipment manufacturers (OEMs) like John Deere or Harley Davidson. But, when was the last time your customer was surprised with a price break or a lead-time reduction? Six Sigma is the way to provide exciting quality as well as improved performance if innovation is allowed to happen. Sometimes, a company can get too caught up in the mundane process of merely controlling rather than improving.

Some consultants and authors might place the Kano model under the measure portion of the Define, Measure, Analyze, Improve, and Control (DMAIC) cycle, but it really helps define the needs of the customer. The Kano model is summarized in Table 4-1.

Table 4-1. Kano model summary

Relationship to Six Sigma	The Kano model helps define what would really make the customer happy.
Who needs and uses it	Teams developing the Define stage of Six Sigma
Cost	Low
Strengths	Helps teams think outside the box
Limitations	The approach is subjective—customers are all different; what's important to one may not be to another.
Process complexity	Simple
Implementation time	1–4 weeks
Additional resources	See Bibliography
Internet search key words	Kano model
Internet URLs	deming.eng.clemson.edu/pub/tqmbbs/prin-pract/perspect.txt www.triz-journal.com/archives/1999/10/e/ deming.eng.clemson.edu/den/archive/98.05/msg00109.html www.imibiz.com/market_r1d.html

QUALITY FUNCTION DEPLOYMENT

Back in 1983, a production manager in a precision sheet metal shop had a sign over his desk that read: "Do you want it fast, or do you want it right?" It was meant to be funny, but the message had serious undertones. If you want one thing, do you have to give up something else to get it? Customers find themselves in this position whenever jobs are quoted or designed for them. They often have to determine the relative importance of one feature against another. From this common reality in every supplier/customer relationship sprang the need for Quality Function Deployment (QFD), which is a fancy term for making sure that the decision will be a good one, based on sound business judgments rather than impulsive decision making.

You would think that in smaller Mom-and-Pop shops, the techniques of QFD probably seem to happen almost by default. Regardless of company size, the distance between what the customer wants and what marketing interprets—or between what the design engineer draws and what the production process generates—can become like the old game of whispering into one person's ear and having them pass the message on. The end result is seldom what the initiator intended. QFD is a methodology used to resynchronize the critical connection between the customer and provider.

QFD focuses on the needs and Voice of the Customer. "Right-the-first-time" could define the desired outcome of the QFD process. Ford and General Motors were among the first US companies to apply QFD tools after seeing the positive effects they had in the shipyards of Japan and later at Toyota. QFD is a customer-driven process for planning the manufacture of products or the delivery of services.

Made visual through the use of a matrix called the "House of Quality," QFD forces all participants to examine the real needs and conflicts in the process. QFD matrices vary widely, but they all try to demonstrate visually the interrelationship between specifications, targets, capabilities, and priorities. The House of Quality is a comparison of what is important to the customer and how a company can match that need or want (Figure 4-2).

There can be up to a dozen steps to filling in the House of Quality graphic. Although some projects may have more or less steps,

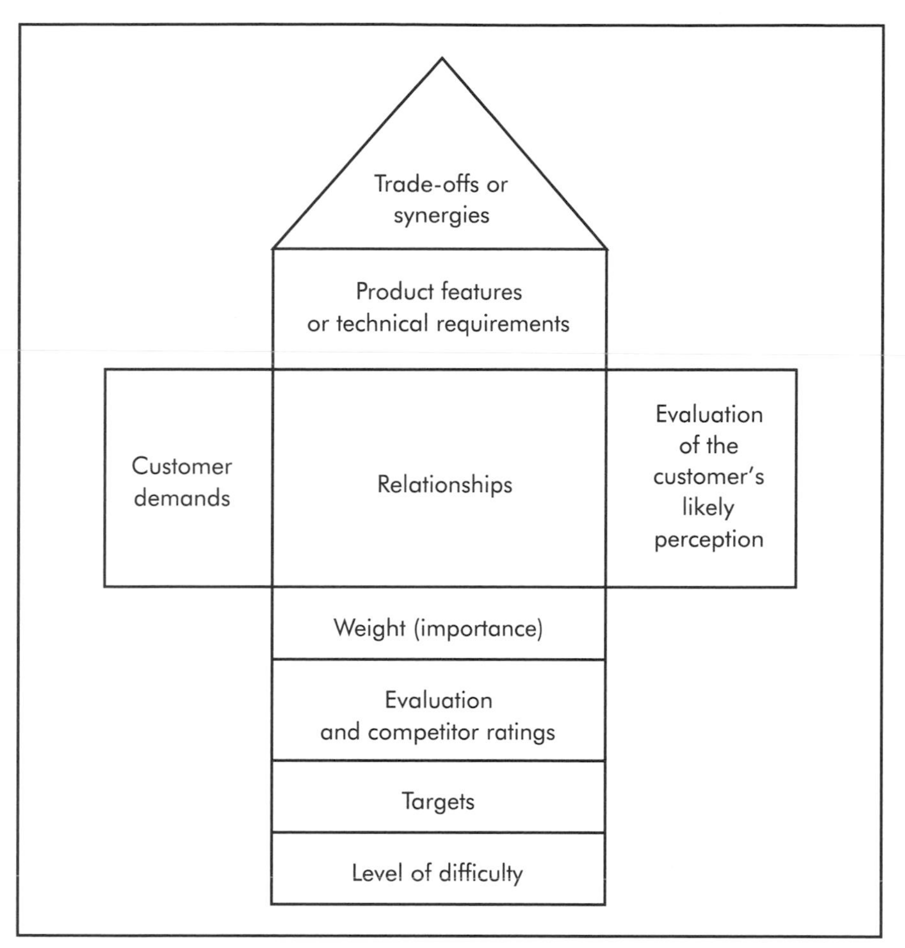

Figure 4-2. House-of-Quality concept.

all steps must be evaluated to complete the House of Quality. The steps are:

1. Identify customer/non-customer requirements.
2. Prioritize/establish an importance rating.
3. Define technical design specifications.
4. Fill out a relationship matrix.
5. Evaluate customer perception.
6. Add competitor(s) product ratings.
7. Identify target rating.

8. Identify level of difficulty.
9. Weight importance to customer.
10. Weight importance to design requirement.
11. Fill in trade-offs/synergies.

The tradeoffs, located in the roof of the House of Quality, indicate the synergistic or detrimental impacts of changes in design. This area is used to identify possible compromises in the design. Since compromises are likely to be encountered sooner or later, they should be explored early on so that potential changes to the design are considered and costed as soon as possible.

There are often conflicts (see also Theory of Constraints) between what the customer wants (for example, good quality, yet cost effective) and design features (for example, strong, yet lightweight). There also may be conflicts in developing the manufacturing process itself (for example, quick set-ups without expensive hard tooling).

Where there is a correlation or synergy between elements (where a positive relationship exists between design or technical requirements), the House of Quality will make it visible. Similarly, it also makes conflicts obvious.

In the example in Table 4-2, a supplier is trying to match a customer's need for paint with the right paint choice. There are strong positive and negative relationships here that help translate the customer's needs into a final decision by weighing or quantifying those needs. For example, if every strong relationship is equal to nine and the weight of importance assigned by the customer is a three, then their product (9×3) equals 27 points in the importance field at the bottom.

Concurrent or simultaneous engineering and TRIZ (theory of inventive problem solving) are terms that can strongly relate to this approach and are discussed later in this book. Normally the process of design happens long before the manufacturing people are involved. Using the QFD approach to overlap the functions of product development, market research, product engineering, process engineering, and production planning will help focus on the design and its manufacturability as well as measure how closely it will provide the customer with what they need or really want.

Defining what the customer wants or needs is the first step. The customer may not even know what they want or what is avail-

Table 4-2. Simplified House-of-Quality matrix

● = Strong = 9 ❍ = Medium = 3 ■ = Weak = 1 Blank = Neutral **Customer Requirements ↓**	Importance to customer	**Technical Requirements ↓** Low material cost	Ease of application	Color choices	Ease of clean up	Ease of disposal
Speed	1	●	●	❍	❍	❍
Coverage	3	■	■	❍	■	■
Durability	3	❍	❍			
Low cost	2	●	❍	■	❍	❍
Importance (weight)		39	27	14	10	10
Target		< \$20/gal (< 5.28/L)	5,000 ft²/day (465 m²/day)	>250	Water based	Sewer

able. Thus, this step includes educating the customer or consumer about what technologies or choices are available.

The financial incentive to capitalize on the tools of QFD is in the reduction of engineering changes. When all the customer concerns have been clearly communicated early in the process—every key question answered and all design conflicts resolved—it is not unusual to see a significant reduction in the number of engineering change orders. It is always easier to make a change on paper rather than in metal. If performed correctly, the process can fairly quickly provide a company with a:

- definition of its market,
- prioritized list of customers and competitors,

- prioritized list of what customers want to see in a product or service,
- prioritized list of what can be done to satisfy the customers' requirements,
- list of design tradeoffs to weigh any necessary compromises,
- realistic set of target values that guarantee customers will be happy, and
- ultimately, a better product or service.

Quality Function Deployment is summarized in Table 4-3.

Table 4-3. Quality Function Deployment (QFD) summary

Relationship to Six Sigma	QFD is critical to the defining stage of the DMAIC cycle.
Who needs and uses it	Sales, marketing, engineering, and production
Cost	Moderate
Strengths	Clearly defines the requirements and conflicts
Limitations	Identifies problems and what needs to be solved, but does not show how to solve
Process complexity	Difficult if a team does not have an experienced facilitator
Implementation time	1–4 weeks, depending on the complexity of the project
Additional resources	See Bibliography
Internet search key words	House of Quality, Quality Function Deployment
Internet URLs	www.ams-inc.com www.qfdi.org www.npd-solutions.com/qfdsteps.htm

SUPPLIERS, INPUTS, PROCESSES, OUTPUTS, AND CUSTOMERS

Part of the define sequence, Suppliers, Inputs, Processes, Outputs, and Customers (SIPOC) is a visual technique used to ensure the entire team understands and looks at the process being examined in the same way.

SIPOC diagrams are relatively easy to complete by following these basic steps:

- Use self-adhesive notes on a wall or flip chart (to make it easier to move things around).
- Title five columns (S-I-P-O-C).
- Begin in the middle by identifying the process.
- Identify the outputs of the process.
- Identify the customers who will receive process outputs.
- Identify the inputs required for the process to function as planned.
- Identify the suppliers of the inputs required by the process.
- Verify the results with the project sponsor, champion, or other stakeholders.

See the SIPOC example in Figure 4-3. Although making coffee seems like a relatively simple process, there is more to it than meets the eye. Once all the inputs that could effect the outputs are defined, even simple processes like this one are complex. Then think how much more complex all the interrelationships and interdependencies are when the task is producing something like a car or computer.

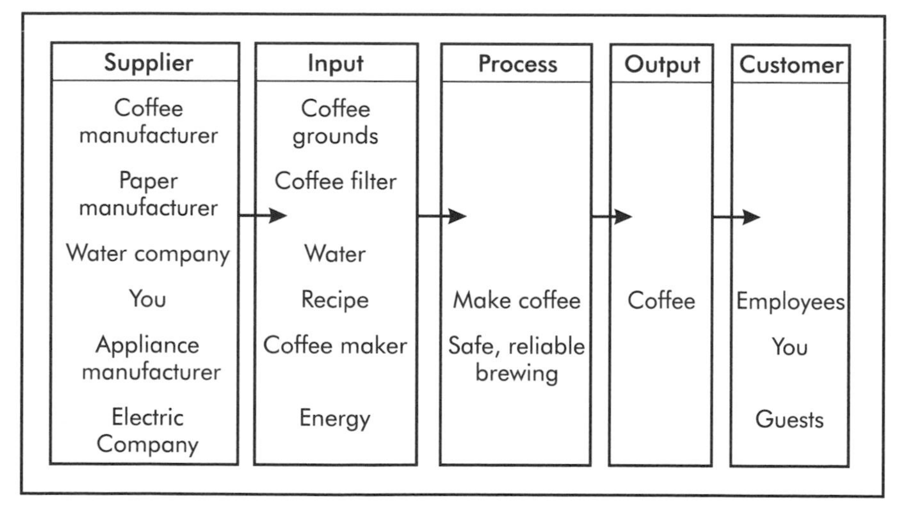

Figure 4-3. SIPOC example.

Many things are taken for granted until they are made visual. You may never think about all the steps in the coffee-making process. You've done it a thousand times; it comes without thinking. Yet when you examine this process, you realize that any step left out, or done in the wrong sequence, can affect the outcome. Figure 4-4 depicts a SIPOC process map for making coffee.

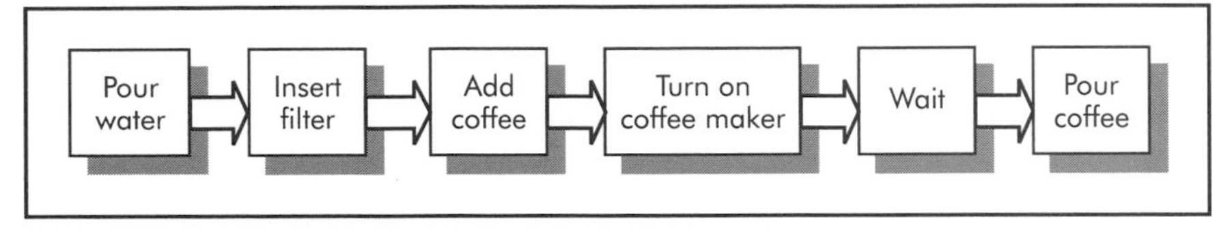

Figure 4-4. SIPOC process map.

Looking at all the other mapping tools (for example, Value Stream Mapping, quality mapping, etc.), it may seem that using the SIPOC approach is redundant. However, some of the other mapping tools focus primarily on the in-house processes that are within a company's control. They do not address all the upstream supplier activities that could make or break the quality of the final product. For this reason, SIPOC is a powerful tool for helping identify early in the process a potential for improvement. One of the rules of Kaizen (continuous improvement) is "understand the process." SIPOC can help a company do that. The approach is summarized in Table 4-4.

ROLLED THROUGHPUT YIELD

Most companies already measure scrap or repair, so yield is an easy calculation. Simply count the number of units lost, divide by the number of good units, and subtract from 100%. For example, if four units are scrapped out of 300, then $100\% - (4 \div 300) = 98.67\%$. What this does not reveal is how many defects those four defective units had.

Throughput yield allows for a slightly deeper examination into the true output quality. If, for example, those same four defective units each had the same three defects, then there are actually 12 defects, so the throughput-yield formula now becomes: $100\% - (12 \div 300) = 96\%$.

Table 4-4. Suppliers, Inputs, Processes, Outputs, Customers (SIPOC) summary

Relationship to Six Sigma	Helps define the project scope
Who needs and uses it	Steering teams, project managers, black belts, green belts, improvement teams
Cost	Low
Strengths	Assists in making visual what might ordinarily be overlooked
Limitations	Offers no solutions, focuses only on the customer as the single critical consequence
Process complexity	Easy
Implementation time	1–5 days
Additional resources	See Bibliography
Internet search key words	SIPOC, process mapping, Six-Sigma tools
Internet URLs	www.qualitydigest.com www.smartdraw.com/specials/isoflowcharting.asp?id=15242

Rolled throughput yield takes this one step further and calculates the aggregate (total) number of defects across all the processes to find the true measure of total yield. For example, if the process described earlier is next to another also producing 96% yield, and another process that yields 99%, then the rolled-throughput-yield formula would stack (multiply) these three throughput yields: $0.96 \times 0.96 \times 0.99 = 0.9123$ (or 91.2%). Obviously, there are some opportunities here for improvement if a company is generating nearly 9% defective product. Keep in mind this is not saying that a company would have 9% scrap. The number of pieces with defects may be only a few percent, but rolled throughput yield counts every defect rather than just parts that have a defect. Rolled throughput yield is covered in Table 4-5.

VOICE OF THE CUSTOMER

Although most companies claim that customer satisfaction is at the top of their list of key initiatives, it has been discovered

Table 4-5. Rolled throughput yield summary

Relationship to Six Sigma	Helps establish where effort should be applied in the planning stages
Who needs and uses it	Black belts, value-stream leaders
Cost	Low
Strengths	Helps identify true quality costs
Limitations	Does not fix anything, simply identifies areas of concern
Process complexity	Easy
Implementation time	1 week
Additional resources	See Bibliography
Internet search key words	Rolled throughput yield
Internet URLs	www.qualitydigest.com www.asqpgh.org/6stools.htm

that a very small percentage (less than 25%) use any method to accurately measure customer satisfaction.

Designed to help measure and define the customer requirements, needs, expectations and perceptions, the Voice of the Customer (VOC) tool is important to all parties involved in the marketing, sales, manufacture, distribution, and service of materials to the customer. In other words, everyone in the organization should understand what the customer expects. Understanding comes from using the DMAIC process to:

- Define who the customer is, what they need, and what the company needs to know to satisfy them.
- Measure by collecting customer opinions, feedback, and complaints, through phone calls, surveys, and feedback sheets.
- Analyze by distilling the feedback into a list of customer needs (expressed in their language).
- Improve by translating customer needs into Critical-to-Quality (CTQ) measurements. Brainstorm ideas to ensure that these CTQ elements are visible and known to those able to control these features.
- Control by establishing specifications for each CTQ feature and auditing the process(es) to ensure compliance.

Defining a list of wants and needs along with the satisfaction/dissatisfaction levels of customers is necessary before using the Quality Function Deployment (QFD) system described previously.

Everyone is a consumer, and when you go to the store to buy a product, you have certain expectations. If your expectations are not met, you are not satisfied. Even if you are given more than you wanted for the same price, you may not be more satisfied. There may be other things that will increase your level of satisfaction. VOC is about finding out what makes customers happy or unhappy.

There is an old saying: "A happy customer will tell three other people, an unhappy one will tell 20." The goal in using the VOC process is to find out what is important to the customer. As a supplier, a company may promote a new product feature, when in reality the customer may be unsatisfied with the change. For example, your morning coffee needs to be hot, yet not too hot. If the coffee is not hot enough, it gets cold before you can drink it all. Too hot, and you could burn your tongue and be unable to taste anything for 72 hours. What is the perfect temperature? As a customer, only you can tell when it is too hot or cold. Whose job is it to measure the temperature? It is the job of the coffee seller. You would be a more satisfied customer if your coffee provider understood that this feature is important to you.

In contrast, coffee vendors have identified as important another feature that may not be important at all. They put a chocolate-covered coffee bean on top of your coffee cup, thinking this "bonus" will make you a happy customer. In reality, you may not like it at all. It falls off the cup while you're trying to make change, bounces down, rolls around between the seats, and can melt and stick to the seat-adjustment rail. So this does not make you more satisfied. It dissatisfies you.

The sample VOC survey shown in Figure 4-5 illustrates how an oil-change company solicited customer feedback on issues critical to quality.

The next step is to turn the responses into meaningful tools for making decisions. Visual tools like those shown in Figures 4-6 and 4-7 may be preferred.

The goal is not just to ask, but to listen and develop a keen awareness of what is really important to the customer. In the shop,

Thank you for taking time to complete this Rapid Oil survey. We appreciate your feedback.

Please check the appropriate box.

Gender	**Age Group**
❑ Male	❑ 18–25
❑ Female	❑ 26–40
	❑ 41–50
	❑ 50 +

Please rate how important and how satisfied you were with these characteristics during your visit.

Importance	**Satisfaction**
1 = Unnecessary	1 = Very dissatisfied
2 = Not important	2 = Dissatisfied
3 = Neutral	3 = Neutral
4 = Important	4 = Satisfied
5 = Absolutely important	5 = Very satisfied

Facilities

Features (popcorn, coffee, etc.)

1❑ 2❑ 3❑ 4❑ 5❑ Importance
1❑ 2❑ 3❑ 4❑ 5❑ Satisfaction

Cleanliness

1❑ 2❑ 3❑ 4❑ 5❑ Importance
1❑ 2❑ 3❑ 4❑ 5❑ Satisfaction

Service

Speed

1❑ 2❑ 3❑ 4❑ 5❑ Importance
1❑ 2❑ 3❑ 4❑ 5❑ Satisfaction

Friendly staff

1❑ 2❑ 3❑ 4❑ 5❑ Importance
1❑ 2❑ 3❑ 4❑ 5❑ Satisfaction

Products

Brand names

1❑ 2❑ 3❑ 4❑ 5❑ Importance
1❑ 2❑ 3❑ 4❑ 5❑ Satisfaction

Prices

1❑ 2❑ 3❑ 4❑ 5❑ Importance
1❑ 2❑ 3❑ 4❑ 5❑ Satisfaction

What suggestions do you have for improvement?

___

___

___

Figure 4-5. Sample oil-change company customer survey.

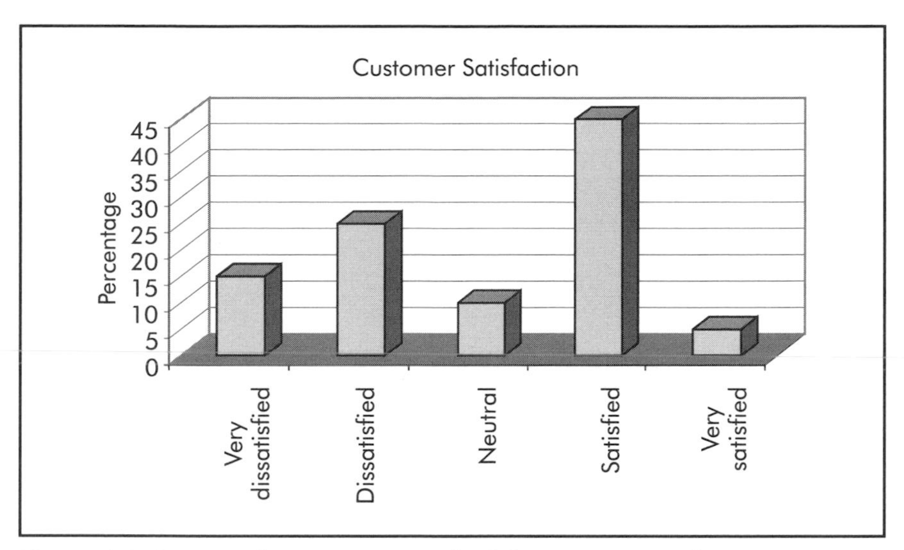

Figure 4-6. Bar-graph presentation of VOC customer-satisfaction survey.

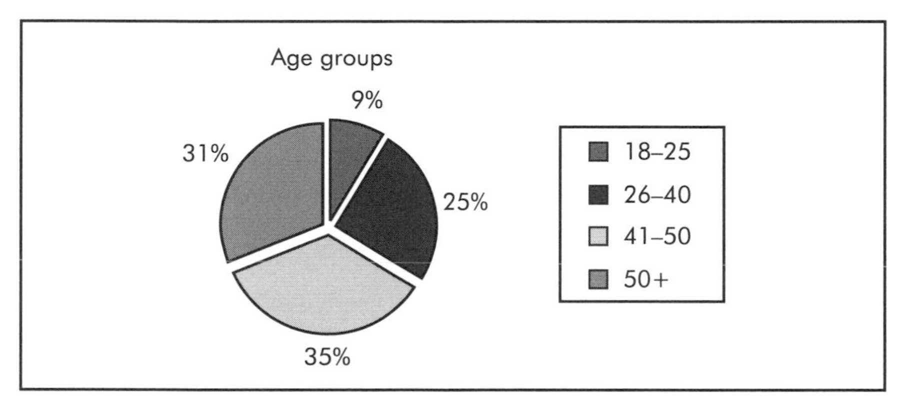

Figure 4-7. Pie chart of age-group response.

this may require talking to the next process, or the next department. Engineering is the service provider to the shop, just as marketing is to engineering.

The message is, listen to the customer. If a company doesn't take the time to ask, or if it doesn't listen, the customer probably will not express that he or she is not satisfied. If he or she is an external customer, they just quit showing up. VOC is summarized in Table 4-6.

Table 4-6. Voice of the Customer (VOC) summary

Relationship to Six Sigma	Critical tool in defining where improvements can be made
Who needs and uses it	Marketing, design, manufacturing
Cost	Low to moderate
Strengths	Lets the customer know a company cares enough to ask
Limitations	VOC does not solve anything. It focuses only on the customer and ignores the balance sheet.
Process complexity	Moderate
Implementation time	1–4 months
Additional resources	See Bibliography
Internet search key words	Voice of the Customer
Internet URLs	www.ams-inc.com www.voicecustomer.com

CRITICAL-TO-QUALITY TREE

Following closely on the heels of VOC, the Critical-to-Quality (CTQ) tree is a visualization tool that can help identify customers' specific needs and drivers. You may not always need to perform this step, but an example of when it may be appropriate or important is when customer expectations are hard to categorize.

There are times when a customer requirement is really unspecified, or so broad that it is difficult to categorize or define in terms of a measurable need. The CTQ tree helps a team make the transition from a macro level view to micro level examination of features that may be critical to quality.

Figure 4-8 shows a simple example of a CTQ tree for the always vague and hard-to-define need called "good customer service." This example is for a call center where technicians deal with computer-related technical questions.

Critical-to-Quality items are the key measurable characteristics of a product or process where there are known performance standards or specifications that must be met to satisfy the customer. These characteristics are defined as important or critical

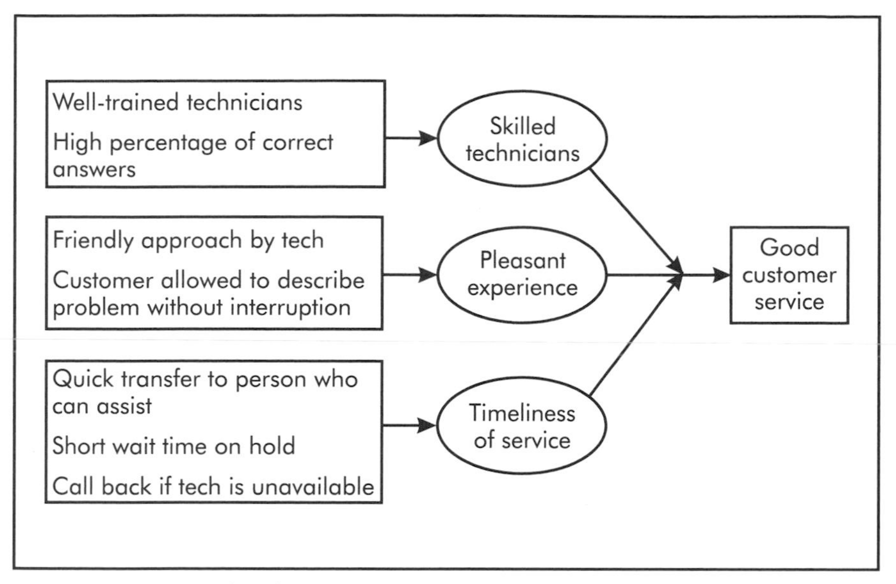

Figure 4-8. Example of Critical-to-Quality (CTQ) tree.

by the customer. Again, this can be an internal or external customer. In layman's terms, CTQ characteristics are what the customer expects of a product. Because the customer often expresses these in subjective language, they must be converted into something measurable. The CTQ tree is summarized in Table 4-7.

AFFINITY DIAGRAM

First, the word "affinity" must be defined in the context it is used here. If you look it up in a dictionary or thesaurus, you will find it alongside words like resemblance, alliance, similarity, relationship, or kinship. Affinity diagrams are really about putting things together that are alike.

At times, the data-collection process generates so much data about so many issues that it can be overwhelming. An affinity diagram allows the team to sort out issues into categories, and then attack clusters of like issues. Teams may also end up dealing with issues that are more qualitative (based on a judgment) rather than quantitative (based on a measurement). An affinity diagram (Figure 4-9) is a means to help sort out the trees from the forest.

Table 4-7. Critical-to-Quality (CTQ) tree summary

Relationship to Six Sigma	A CTQ tree establishes the elements important to the customer. It helps define critical information.
Who needs and uses it	Sales and marketing teams, project managers, black belts, teams
Cost	Low
Strengths	Helps team verbalize what might otherwise be overlooked
Limitations	Does not offer solutions
Process complexity	Easy
Implementation time	1–5 days
Additional resources	See Bibliography
Internet search key words	CTQ, critical to quality, Six-Sigma techniques
Internet URLs	www.sixsigmaforum.com/concepts/c2q/index.shtml www.ge.com/sixsigma

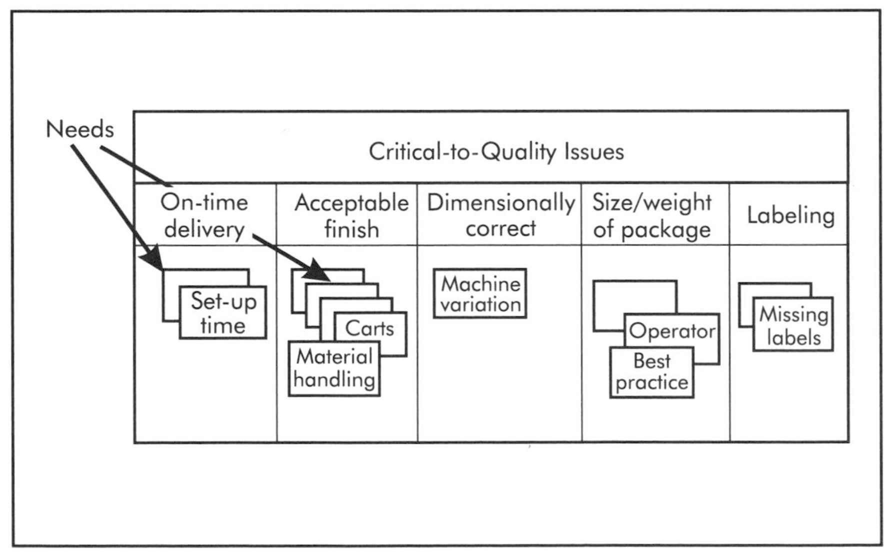

Figure 4-9. Affinity diagram.

The steps to developing a meaningful affinity diagram are:

- Form a cross-functional (diverse) team of less than 10 participants (5–9 works best).
- State the issue or problem to the team (using a flip chart or white board).
- Provide the team with self-adhesive notes for writing ideas down (one idea per note).
- Give each participant about 15 minutes to write down three to five causes or ideas to solve the problem (in a sentence format).
- Collect all the ideas and spread them out on a table or wall randomly.
- Ask participants to silently group ideas (loner ideas are kept separate).
- Review groupings and then title each group.
- Determine which note within each group captures the essence of that group, and use this as a header or create a header.
- Send the team back to try to generate more ideas (since this activity usually spawns more creative thinking).
- Develop an affinity diagram by organizing groups that are closely related close to each other.
- Review and modify titles if needed.

As the team categorizes the statements and brainstorms issues into clusters, participants should try to perform this step silently, with each team member feeling free to move a statement if he or she thinks it is placed in the wrong category. The affinity-diagram approach is summarized in Table 4-8.

MANAGEMENT BY WALKING AROUND

Politicians would put it like this: "Get out and press the flesh." To find areas for improvement, walk around and interface with the people who are the closest to the customer or the hourly workers on the shop floor. If Six Sigma is going to work, and if management is going to be viewed as supporting the process for change, managers have to be viewed as interested in change, and interested at every level. This requires them to be visible and curious about what is going on.

Table 4-8. Affinity diagram approach summary

Relationship to Six Sigma	Assists in brainstorming potential solutions and improves potential for a breakthrough
Who needs and uses it	Engineers, marketing, administrative, and production functions, team facilitators
Cost	Low
Strengths	Challenges teams to think outside the box and helps organize ideas into natural groupings
Limitations	Can be too subjective and is not often based on hard data
Process complexity	Easy
Implementation time	1–2 days
Additional resources	See Bibliography
Internet search key words	Affinity, TQM tools, Six-Sigma techniques
Internet URLs	st-div.web.cern.ch/st-div/qa/minutes/ QWG-990617/affin.html mot.vuse.vanderbilt.edu/mt322/library.htm

Popularized in a book by Tom Peters and Robert Waterman, Management By Walking (or wandering) Around (MBWA) was developed by executives at Hewlett-Packard in the 1970s. Top managers who routinely wander around outside the normal confines of their cubicles or offices have a much greater opportunity to engage employees in meaningful interaction.

Increasing communication with suppliers and customers through MBWA has also proven more successful for many managers than operating in a more isolated environment. By allowing for informal communication and a reduction in the bureaucratic lines of communication, MBWA promotes more effective transfer of organizational values and philosophy more frequently and directly on a one-on-one basis.

The key to effective MBWA is not to just put miles on your shoes. To be truly beneficial, the manager must be walking around with his or her eyes wide open. He or she should be observant and ask questions. The greatest obstacle with MBWA is that managers are not born effective wanderers. Managers, through their training and experience, are more often tellers rather than listeners. It is the listening that makes wandering so beneficial. Being a good questioner is the secret. This is not meant to be simply a social visit. This interruption of an associate's day must be meaningful. Quality rather than quantity is the watchword when wandering. One or two discoveries about which a manager can quickly show a response is probably of greater value than creating a list of 25 action items that will probably be overwhelming and never get the attention they may deserve.

MBWA allows managers to provide real-time feedback and point-of-use assistance. People are not generally inclined to speak up in public. Bring 20 people into the lunch room and ask for ideas for improvement or ideas about what might be creating a problem, and you'll probably get blank stares. Instead, go directly to the work area. The issues that operators deal with every day are right there in front of them. One on one, you'll probably get an earful. So be ready to listen. Other employees will also see that you get around and are willing to listen, and they will mentally prepare themselves for your visit.

In nearly every case where teams struggle, there is a direct correlation between the level of support they have received in the past and the amount of effort the team is willing to give during any new initiative.

My most stressful training experience was in Chippewa Falls, Wisconsin, in the middle of winter seven years ago. I was sent in to train the last 40 people out of a company of 180. To the management teams' credit they saw the need to train. This was my first visit to the company, and to my dismay, I found that they had saved the malcontents for last.

The four-day workshop started off with me talking over the tops of 40 heads. There was no eye contact, no engagement, and little if any response to any of my questions or efforts to draw them into a conversation. At the first break, I tried to be very polite with a couple of them, to get a relationship going, and have a better chance at breaking the log jam loose. No chance.

I told my best joke after lunch. Nothing. I realized then how a pig at a barbeque must feel. Somehow, I struggled through the rest of that day, and that night seriously questioned my calling as a trainer.

At the training session the next morning, I pulled up a chair into the middle of the U-shaped workshop area, sat down, and quietly but matter-of-factly explained that being 1,800 miles from home, away from my family, was hard enough already. I made it clear that I was unwilling to face another day like the one before. If they were unwilling or unable to participate, I said I would be happy to gather my equipment and catch the next flight home, not because I wanted to, but because I was not willing to waste the company's time and money.

After what seemed like 30 minutes but must have been less than 30 seconds, someone finally broke the awful silence. "We know that what you're talking about (Lean Manufacturing) works. It just won't work here." I asked him why he felt that way. The log jam had broken. Everybody started talking at once. I had to jump up to get a pen and paper to write down their comments, because there was no way that I was going to remember all of the remarks and sources of frustration they were identifying.

Once it quieted down a little and people had a chance to decompress, I asked if I could transfer all of their comments to a flip chart while they took a break. People kept coming up to me to add more. Before we had reassembled around the tables, we had filled four flip-chart pages of single-spaced lines, with all the reasons why they believed company management would sabotage their efforts.

They made a good case. As it turned out, this two-hour venting session turned into a very productive discussion about how to expand the circles of influence. After explaining Stephen Covey's theory of the circles of concern versus circles of influence, they came to agree that there were certain things they had listed that were outside of their control, things they were concerned about, but over which they had no influence. On the other hand, there were also items that they did control.

Systematically we drew lines through the items on the list that they viewed as outside their control, and this left about a dozen items (out of the original 40). These were items that they felt were within their circle of influence. They decided as a group that setting aside the things outside their control and focusing on the items within their control would help them concentrate on where such improvement was truly a possibility.

By choosing to participate, they agreed that there was indeed a greater opportunity to expand their circle of influence rather than collapse it by simply refusing to be engaged. Instead of withdrawing and hoping the initiative for change would just go away, they were saying that they recognized that by failing to participate, they were giving away their right to control part of their own destiny within the company.

The team agreed to participate, and they ended up being one of the most excited group of machinists I have ever worked with. Later, I had a chance to facilitate a couple of very successful Kaizen teams with members of this group (to reduce set-ups on machines that were previously a significant bottleneck). The gains recognized from their efforts equaled hundreds of thousands of dollars and massive reductions in machine set-up time that translated into major lead-time reductions.

The point of the story is that all the stress and all the time wasted trying to motivate a group of seasoned machinists could have been avoided if they had greater confidence in management's commitment to the new initiative. People have a short memory, except for your mistakes. Don't let the memories of mistakes from long ago defeat future efforts. If you need to bury some baggage, do it. Let people know that you recognize the problems, let them vent, and then establish those items that are within your control to remedy. Don't overpromise and underdeliver.

Set aggressive, yet attainable goals. Let the teams know it cannot be done without their help. Every day each of us must wake up and look in the mirror and remind ourselves that we must focus on those things that we can control. You can also help others do the same. If you remove those concerns about which you have no control from your main field of view, you can accomplish great things by focusing on the issues that you really can effect.

Management By Walking Around is summarized in Table 4-9.

HOSHIN KANRI

As you might expect, Hoshin Kanri originally developed in Japan. It is being adopted by many Western companies in an effort to bring a more strategic focus to planning and executing daily tasks. Gaining popularity in recent years, it has become part and parcel with the Six-Sigma toolbox.

Table 4-9. Management By Walking Around (MBWA) summary

Relationship to Six Sigma	Helps management find out what is in the heart of the people at all levels
Who needs and uses it	Managers, team leaders
Cost	Low
Strengths	Allows managers to keep their hand on the pulse of the organization
Limitations	Does not fix anything on its own
Process complexity	Easy
Implementation time	From first day onward
Additional resources	See Bibliography
Internet search key words	MBWA, Management By Walking Around
Internet URLs	www.kalikow.com/~drdan/mbwa-txt.shtml www.lib.msu.edu/lorenze1/mbwa.htm

Hoshin is often translated as policy, but vision may be a more descriptive term. Hoshin more literally refers to a compass or pointing device. *Kanri* means management or control, and when used together with Hoshin, these terms have been conceptualized as compasses directing individual ships as they move in concert with a larger fleet. Even if the individual ships are surrounded by fog or shrouded in darkness, they can all move together by reading a common compass. The same is true for companies needing to guide daily activities and decisions. Working together toward a common goal through the use of a common pointing device can save untold effort and enormous amounts of time and money.

For years, management by objectives was the preferred approach in America. Everybody was basically on their own, like one ship captain trying to guess what all the other ship captains were doing. If each ship captain is measured on his or her own, and against a unique set of objectives, then naturally the captains will each seek to perform in a manner that makes themselves look good—regardless of whether this supports the goals of the fleet.

Management by policy and use of the following four stages of Hoshin Kanri (the development and execution of policy) make sure all the ship captains work together toward a common goal. The four stages of managing by policy are:

- policy setting,
- policy deployment,
- policy implementation, and
- policy evaluation and feedback.

By ensuring that the individuals (captains) understand that they do not succeed unless the mission of the fleet is successful, there is a greater motivation to work with other teams and departments (ships) rather than focusing on only measurements that make the individual captain look good. This avoids suboptimization.

Hoshin Kanri is a methodology for capturing and translating the organization's vision and objectives into executable, actionable, and measurable strategies for the company. By documenting and vocalizing (writing down) a meaningful purpose, the long-term direction for the company is less likely to evaporate or be diluted by day-to-day firefighting or passive resistance.

Other names for Hoshin Kanri include policy management and policy deployment. One of policy management's greatest strengths (using Hoshin Kanri) is its ability to translate qualitative (subjective) goals into quantitative (objective) actions.

Four key elements of the Hoshin cycle are shown in Figure 4-10, overlayed onto the plan-do-study-act (PDSA) approach commonly attributed to Dr. W. Edwards Deming, and sometimes referred to as the Shewhart cycle.

The four elements in sequence are:

1. Policy setting—this phase focuses on the selection of a vital few objectives, usually no more than four or five.
2. Policy deployment—value-stream managers then translate how their area of responsibility will support the few critical objectives. Similar to a mission statement, a Hoshin is a combination of a statement of a strategy, plus statements of the means to achieve the strategy and how performance will be monitored (sometimes called a target). Area managers develop their Hoshin, and then pass it on to subordinates who will in turn translate their part as a sub-Hoshin, with its own strategy, means, and targets.

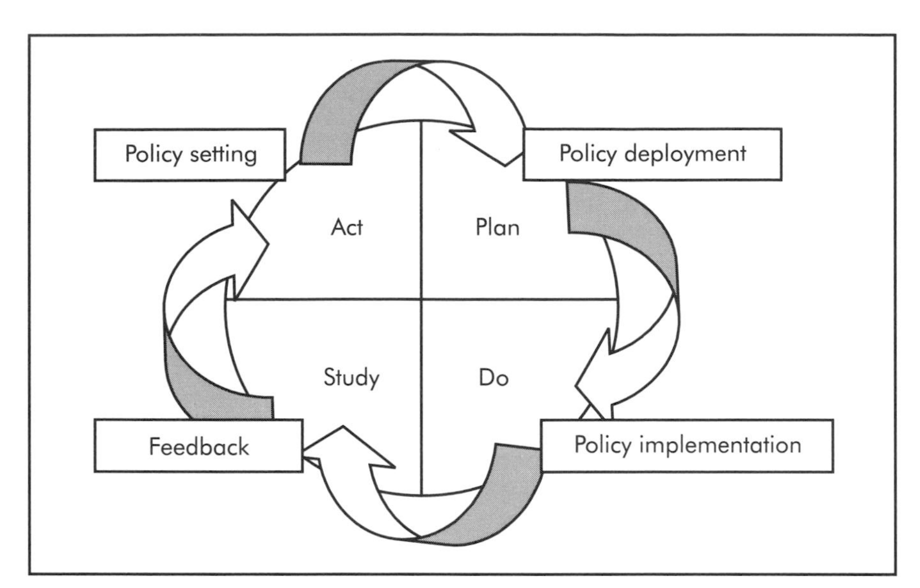

Figure 4-10. PDSA overlaid with Hoshin cycle.

3. Policy implementation—once the Hoshins have been accepted, implementation plans are developed. Gantt charts and other time-phased project-planning tools are often used to provide clear direction, time fences, and areas of responsibility.
4. Policy evaluation—like calibrating the compass, Hoshin targets are routinely reviewed and, if necessary, modified where there is an ongoing problem or need to change direction. Project information is distilled for senior management to review prior to an annual cycle that begins the process all over again.

Use of this four-phased approach is generally spread over one year, beginning with an annual planning activity. The balance of the year is spent executing the plan.

Speaking in terms of the Shewhart (plan-do-study-act) cycle, begin with the act stage where executive staff revises corporate strategies with general managers. This is the focus phase of Hoshin Kanri, where the critical few strategic objectives are determined. Follow this step with the "plan" stage to distill the vital few down to cross-functional objectives within the company so that departmental and value-stream priorities are brought into alignment with the whole "fleet."

The rubber meets the road in the "do" stage. This is where the priorities are integrated into the daily management tasks of individual managers and projects. The "check" stage has to do with managing and measuring key objectives in the process of daily work. This may include daily or periodic audits. A review (audit or evaluation) toward the end of the annual cycle is common. In this feedback or responsiveness phase, progress is monitored. If necessary, course correction or corrective action is taken. Then, the entire process repeats.

Hoshin Kanri is summarized in Table 4-10.

Table 4-10. Hoshin Kanri summary

Relationship to Six Sigma	Helps to keep the Six-Sigma program on track by choosing the right things to work on
Who needs and uses it	Management, value-stream leaders
Cost	Low
Strengths	Helps translate qualitative goals into quantitative actions
Limitations	Does not offer solutions, simply identifies the issues to work on
Process complexity	Moderate
Implementation time	1–4 weeks
Additional resources	See Bibliography
Internet search key words	Hoshin Kanri
Internet URLs	www.mgt.uea.ac.uk/research/witcher-b/bjw1.asp www.tqe.com/hoshin.html

PRODUCT, QUANTITY, ROUTING, SUPPORT SYSTEMS, AND TIME

The product, quantity, routing, support systems, and time (PQRST) approach asks these key questions:

- Product—what is made?
- Quantity—how much is made?
- Routing—where is it made (and with what machines)?
- Support systems—what outside process or shared resource is required?
- Time—how long does it take to make?

A multi-purpose tool, PQRST is useful when selecting a model line, when performing a facilities layout plan, or when management needs to perform the Hoshin-Kanri step.

Having the ability to quickly and accurately define the value streams in an organization and the processes within those value streams can help speed up the planning process, and assist in selecting the correct things to work on. An ancestor of Value Stream Mapping, the PQRST process helps identify where equipment and activities should be located in relationship to other equipment and activities. Like Value Stream Mapping, this activity is best performed once product families have been identified. Yet, the tool itself is a method to help identify product families. So, there is a chicken-before-the-egg question here (and thankfully the answer is not complicated).

There is a need to perform the PQR portion of the analysis twice—first, on a macro level for all part numbers (or part types) moving through the shop. This step identifies and groups similar parts into families of like products. Again, the analysis is performed at the product family (or value-stream) level—a micro-level PQR analysis within each value stream.

The ultimate goal is to divide and conquer, by means of sorting like parts (much like Group Technology) into logical family groups. This process is made easier when a computer is used. If everything has to be sorted on recipe cards (an option), it could take weeks if there are many part numbers.

A simple example of a PQR analysis for a lost-wax investment-casting process is shown in Table 4-11. Notice that the micro detail has not been identified.

Assembling the wax mold could take hours and require many different processes such as cleaning, trimming, welding, and blending. The process happens all in one place and the wax-assembly personnel do not lose control of it until it is ready to move to the

Table 4-11. Product, Quantity, Routing (PQR) analysis for a lost-wax investment-casting process

P	Q	R					
Part Number	Monthly Demand	Mold	Assembly	Invest	Cast	Etch	Clean
101-A	150	1	2	3	4		5
102-A	100	1		2	3	4	5
103-A	100	1		2	3		4
104-A	50	1	2	3	4		5
104-B	50	1	2	3	4		5
105-A	100	1		2	3	4	5
106-A	250	1		2	3	4	5
		Process sequence >					

next process, investing. Patterns here are hard to identify, and the situation is made even worse when multiplied by the hundreds or thousands of active part numbers in the order files. However, through the power of computers and modern query functions, the groups that follow the same path can be quickly sorted, thus assisting in identifying potential families or value streams. Table 4-12 shows the same data, now sorted. This makes it very obvious

Table 4-12. Sorted PQR data

P	Q	R					
Part Number	Monthly Demand	Mold	Assembly	Invest	Cast	Etch	Clean
101-A	150	1	2	3	4		5
104-A	50	1	2	3	4		5
104-B	50	1	2	3	4		5
103-A	100	1		2	3		4
102-A	100	1		2	3	4	5
105-A	100	1		2	3	4	5
106-A	250	1		2	3	4	5

that regardless of customer name, the products are grouped according to process requirements.

There will generally be value streams that share common equipment or a shared resource. The power of knowing these interrelationships will help to position equipment (or shipping locations) to achieve a minimum of handling and transportation.

There is a close relationship between PQR tools and a Pareto analysis of the sales mix. According to Vilfredo Pareto (creator of the Pareto diagram), there is an 80–20 rule that dictates that 20% of the part numbers will make up 80% of the revenue from any given product line or value stream. The power of PQR is only recognized if a shop is set up to capitalize on the 80–20 rule.

There is a tendency, especially in job shops, to be fixated on the need for flexibility (usually required for the remaining 20% part of the business) to deal with "dog-and-cat parts" (those odd-ball parts that make up the smallest percentage of revenue-generating part numbers). Opportunities and facility-layout options are ignored, which might better facilitate the manufacture of those part numbers generating 80% of the revenue.

A company might have to temporarily ignore the interrelationship of activity centers and traffic patterns to initiate change. However, it must avoid long-term optimizing of one cell (or work team) at the expense and risk of suboptimization of the entire operation. PQRST is summarized in Table 4-13.

TEAM CHARTER

The team charter is the rudder for teams. After management or a steering team has determined a need for a Six-Sigma or process-improvement project, they should assist the team by providing them a charter. A charter acts as an agreement or contract between the team and management about what will be accomplished. Project charters are a means to initiate, control, resource (fund), and manage continuous-improvement projects. A clear project charter will help teams avoid dealing with unimportant issues. It reduces overlapping efforts and decreases the chance that the team will spend time working to improve a process that will soon become obsolete. A meaningful project charter helps

Table 4-13. Product, Quality, Routing, Support Systems, Time (PQRST) summary

Relationship to Six Sigma	Helps to define needs and opportunities by identifying distinct value streams
Who needs and uses it	Steering teams
Cost	Low
Strengths	Makes patterns and value-added activities easier to see once similar features are sorted
Limitations	On its own, PQRST offers no solutions.
Process complexity	Moderate
Implementation time	1–4 weeks
Additional resources	See Bibliography
Internet search key word	Process mapping
Internet URL	www.umich.edu/~umjtmp/video.html

define the scope, deliverables, and responsibilities of the team. Figure 4-11 is an example of a team charter.

The team charter's importance to Six Sigma is summarized in Table 4-14.

CONCLUSION

At the end of the define phase, there should be a clear and concise definition of why the selected projects were designated and resourced. Company management should be able to convey why the project is important, along with a definition of the goals and objectives (specifications) for the project. There should be clear ties to the business objectives and vision statement.

Responsible parties should be identified, including sponsors, team leaders, support, and advisor members. Clear scope statements should clarify the processes to be examined and improved. If at all possible, a current-condition statement should define the improvement opportunities for the team. Sacred cows (things that cannot change), resource parameters, and budgets for time also should be defined and delivered to teams in this phase.

Team Charter

Purpose: The X-team will be assigned to reduce the engineering throughput time and improve flat-pattern accuracy.

Problem statement: Engineered drawings currently take between 5–7 days from receipt of customer specifications to the time shop drawings are released. An inordinate amount of time (an average of 15 minutes) is spent by the operators checking the flat-pattern drawings for errors against customer drawings. Initial data shows 25,000 defects per million opportunities (DPMO).

Objective statement: Reduce engineering lead time to 3.5 days and establish a program that will measure this objective against a minimum performance of ± three sigma limits. Improve first-pass quality by reducing it to less than 10,000 DPMO.

Scope: Limited to flat-pattern drawings only

Charter date: September 1, 20XX

Completion target date: October 1, 20XX (see attached project plan)

Stakeholders: Engineering manager, production manager, production supervisors

Sponsor: vice president of operations

Team: black belt, Moe; green belt, Flo; engineer, Joe; punch operator, Zoe; programmer, Barney

Other resources: Engineering workstation

Why is the objective important? 20% of all late shipments and 5% of scrap can be traced to problems related to flat-pattern errors.

Figure 4-11. Team charter.

Table 4-14. Team charter summary

Relationship to Six Sigma	A clearly defined objective helps teams stay on course.
Who needs and uses it	Steering teams, project leaders, black belts, project teams
Cost	Low
Strengths	Avoids false starts, helps teams be self-directing
Limitations	Few
Process complexity	Easy
Implementation time	1–3 weeks, depending on project complexity
Additional resources	See Bibliography
Internet search key words	Team charter
URL	www.ars.usda.gov/afm/tqm/guide lines/wri_cht.htm

5

Step Two: Measure

Taiichi Ohno is credited with saying "Where there is no standard, there can be no Kaizen." Another way of saying this is, "Where nothing is being measured, nothing will be improved." This chapter examines measurement tools, realizing that measuring alone does not improve anything.

Statistics is a powerful medium that converts hard-to-see dimensions into something visual and understandable. There is no way to fully define in this text the hundreds of tools available. Additional resources can be found in the Bibliography. With Value Stream Mapping, spaghetti diagrams, and symbolic dashboards, the discussion of many measurement methods and techniques will be further developed.

A SHORT COURSE IN STATISTICS

The word "statistics" can send chills up the spine of a machine operator. Yet, statistics are used every day: your son's Little League batting average, your vehicle's gas mileage, the average training time for a machine operator, or weekly overtime averages. These are examples of statistics that anyone can understand, not just mathematicians. It is common to see people avoid the use of statistics in areas where the complexity is no greater than in these simple examples, yet the need for them is much more valuable and important.

Any book about Six Sigma must spend a little time talking about the principles and use of statistics in a continuous-improvement program. Statistics are simply numeric descriptions. Measurements help us visualize things that are hard to see. Statistics are a way to build confidence in an observation that is otherwise merely an opinion. They help us measure the performance of one sports team against another or make decisions about which car to buy or where to live.

There are two primary kinds of statistics: descriptive and inferential.

Descriptive Statistics

Descriptive statistics summarize large amounts of data. For example, in a group of 42,341 people attending a football game, 31,656 have valid driver's licenses. Therefore, 75% of all the people at the game were licensed drivers. To get to this degree of accuracy, data would need to be collected for each person.

Inferential Statistics

Inferential statistics use a data sample to make a judgment. For example, if 250 people at the game were interviewed, and 180 were licensed drivers, we could make a judgment or inference that 72% of all attendees were licensed drivers. This inferential statistic is less accurate than interviewing 100% of the attendees, but it saves a lot of time and effort. In this case, the inferential result is 96% accurate compared to the descriptive result, and 4% of the licensed drivers are unaccounted for. A measure of accuracy is given up when using the sampling method to make judgments.

Data

There are many kinds of data used to establish and analyze statistical information, including nominal, ordinal, interval, and ratio data. *Nominal data* classifies data into logical groups. For example, you count 100 passenger vehicles that pass your house to determine the percentage of each category of vehicle (such as, 35 autos, 25 trucks, and 40 SUVs). *Ordinal data* assigns a value or measurement to a sample. For example, you rate the value of each vehicle (for example, greater or less than $10,000 value) as it passes. *Interval data* makes a comparison between any two samples. For example, you measure the amount of time between cars passing your house. *Ratio data* determines how many times one sample is different from another sample. For example, you count the number of people in each car and find the number of times there is more than one person in a vehicle.

Terminology

There are also some key terms in statistics used to aid in a common understanding of the tools, such as population, variable, sample, qualitative, quantitative, mean, median, range, and sample variation and deviation.

A *population* is any set of numbers. For example, all red cars, or all cars with their windows down. A *variable* is the individual property in a population that sets it apart from the rest. For example, any red car that is also a convertible. A *sample* is a smaller subset of a larger population. For example, instead of watching 100 cars pass your house, you may measure a sample of 10.

Qualitative data is judgmental data that is hard to measure. For example, how many cars would you consider clean? *Quantitative* is a measurable feature. For example, all the cars that have 15 in. (38 cm) wheels.

The *mean* is the average value of a population or set of data. For example, the mean (average) of the values 5, 4, 5, 4, and 6 is 4.8. Add all the values ($5 + 4 + 5 + 4 + 6 = 24$) and divide by the number of individual values (5 data points). Thus, $24 \div 5 = 4.8$. The *median* is the middle number in any series of values. For example, arrange the values from smallest to greatest in the series: 4, 4, 5, 5, 6, and find the center number: 5. The center number here is easy to find with an odd number of values, but if you have an even number of values, then the two middle numbers act as the median. The *range* is the difference between the smallest and largest value. For example, subtract the lowest number from the highest number in the series 4, 4, 5, 5, 6 ($6 - 4 = 2$), thus the range is 2. Range is the simplest calculation of variation in measuring a process. Because all of Six Sigma is based on reducing unwanted variation, range is very important.

Sample variance is the sum (total) of the squared distance from the mean divided by the total number of data points minus one (see Sigma calculations in Chapter 2 and Table 5-1).

Turning measurements into a sigma limit is very similar to finding the sample variance in a set of numbers. For example,

Table 5-1. Sample variance

X Entry Data	$\bar{X}$ Mean	Variation from Mean	Squared Value of Variation from Mean
4	4.8	–0.8	0.64
4	4.8	–0.8	0.64
5	4.8	0.2	0.04
5	4.8	0.2	0.04
6	4.8	1.2	1.44
Sample variance			**0.70**

$$(4 - 4.8)^2 + (4 - 4.8)^2 + (5 - 4.8)^2 + (5 - 4.8)^2 + (6 - 4.8) = 2.8$$

$$5 \text{ (total data points)} - 1 = 4$$

So, $2.8 \div 4 = 0.7$

The *sample standard deviation* is the positive square root of the sample variance. For example, the sample variance was just calculated as 0.7. The square root of this number is the sample standard deviation (S). Thus,

$$S = \sqrt{X - \bar{X}} = \sqrt{0.70} = 0.837 \qquad (5\text{-}1)$$

Control Limits

This section will not discuss the process for calculating a control limit (there are many kinds of control limits). Instead, it will focus on the relationship between a target and the control implied by the Six-Sigma limits overlaid onto that target.

When overlaid onto a control chart, sigma limits are useful, much like control limits on a traditional X-bar and R chart. Control limits are values calculated differently than sigma limits, but the resulting values can be very close to the value of the third sigma limit ($\pm 3\sigma$). Like the painted fog lines on a highway, control limits are there to indicate that part measurements are staying within expected values. If the part measurements begin to drift toward one end, either the upper control limit (UCL), or the lower control limit (LCL), then a course correction can be taken. This is not to say you should spend all your time adjusting the process, but seeing trends happen can avoid a crash, whether driving a car or running a process. Trends are evidenced by a number of parts moving without interruption toward one end of a control limit or sigma limit. (If your car were drifting without interruption toward the ditch, you or your passengers would snap to attention and demand a course correction.) Some minor variation back and forth between these lines on the control chart is expected and considered normal. It is the sudden movement or the steady drift toward a limit (the ditch) that is alarming and needs to be looked into and corrected.

Summary

In summary, here is a relevant inferential statistic. A large number of all the people working in your shop probably do not enjoy performing mathematical functions. Yet, being able to manipulate numbers and turn the results into understandable summaries will be a critical skill in applying the Six-Sigma approach. (If you have not practiced math in some time or used statistics, check your local used-book store. Books on math and statistics are among the first sent to the recycling bins.) Thankfully there are new software tools that can help speed up the calculation process and reduce some of the need for training team members. However, some fundamental background is needed for team members to be able to select the right tool at the right time. Otherwise, it's like having a fully stocked toolbox without any mechanical training. They must understand each tool's basic use and when and where to apply it. They do not need a degree in math or statistics to use these basis tools, but the need for some basic understanding cannot be ignored.

For more information on the application of statistics, visit www.statsdirect.com.

STATISTICAL PROCESS CONTROL

With a basis in statistics, you are now ready to see how this math is applied to statistical process control (SPC).

Terminology

Histograms are graphic representations of the "history" of a process. Much like the cement blocks example in the discussion of Six Sigma, the histogram can identify whether the process is producing parts that fall into a normal distribution as shown in Figure 5-1.

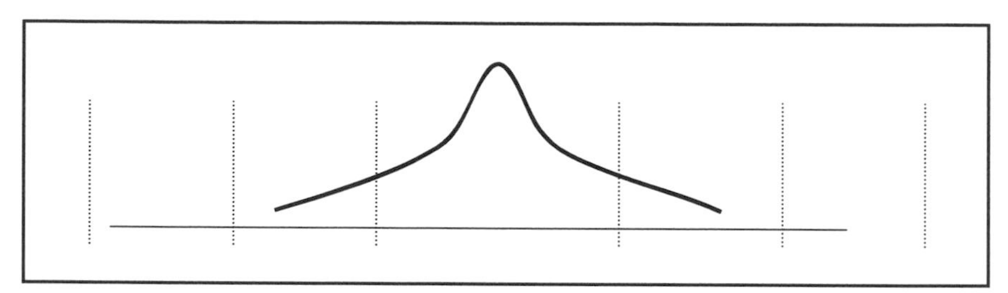

Figure 5-1. Normal distribution.

Not all distributions are "normal." Non-normal distributions have multiple causes. In the cement-block application, this could be caused by having different operators, shifts, or machines producing parts or even two different measurement devices, resulting in measurements showing two peaks (Figure 5-2). This is referred to as a *bi-modal distribution*.

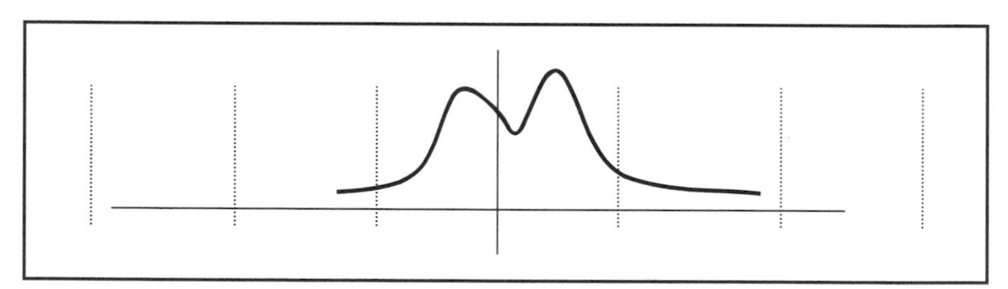

Figure 5-2. Bi-modal distribution.

If bad parts are being sorted out, or if the process allows parts to be larger but not smaller than a standard, a histogram that looks like the truncated or skewed one in Figure 5-3 could be the result. There are dozens of other distribution shapes and types, each with their own explanation of why they look the way they do.

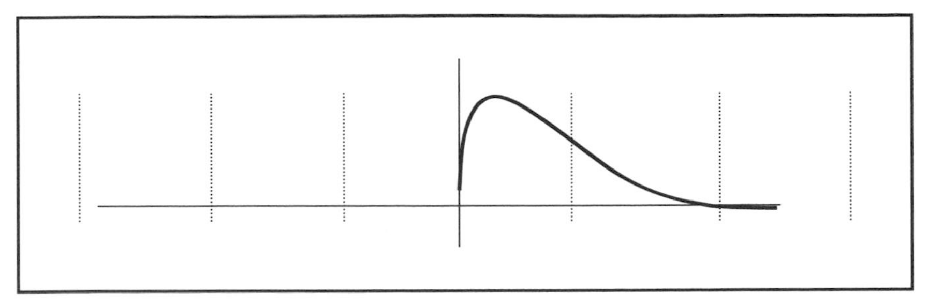

Figure 5-3. Truncated distribution.

X-bar and R Charts

X-bar and R charts are key statistical tools for recording variation. *X-bar* is the mathematical symbol for average ($\bar{X}$) and half the namesake for a two-part chart known as the *X-bar and R chart*. (As discussed earlier, *R* stands for range or the difference between the smallest and largest measurements.)

For example, if five cement blocks are weighed and the results are: 18, 17, 18, 16, and 19 lb, the total weight for all five blocks is 88 lb. To find the average, divide the total (88 lb) by the number of blocks (5) and the result is an average (or X-bar) of 17.6 lb. This average is shown as a line on the upper X-bar chart in Table 5-2. It is not the target, but rather the average in relation to the target.

The range is found by subtracting the weight of the heaviest block (19 lb) from the weight of the lightest block (16 lb). The result is 3 lb. On a chart, the range value can never be less than zero. Plotting these two values (X-bar = 17.6 and R = 3) is the first step in developing a table like the one shown in Table 5-2. Succeeding columns allow you to graphically describe subsequent weighing.

The value of the X-bar and R chart is that it represents the performance of a process over time and helps identify potential problems or trends before they result in a defect. For example, in Table 5-3 even though the range is fairly stable, the average (X-bar) for concrete-block weight is slowly trending downward. Knowing this can prompt a team to find out what could be causing the trend before it results in a crash. On short runs, individual measurements may be plotted. On longer production runs, five or more measurements may be averaged together as in this example.

Table 5-2. Simplified X-bar and R chart

X-bar																			
26																			
25																			
24																			
23																			
22																			
21																			Target
20																			
19																			
18																			
17	x																		$\bar{X}$
16																			
15																			
14																			
13																			

Range																		
8																		
7																		
6																		
5																		
4																		
3	x																	
2																		
1																		
0																		

Table 5-3. Simplified X-bar and R chart

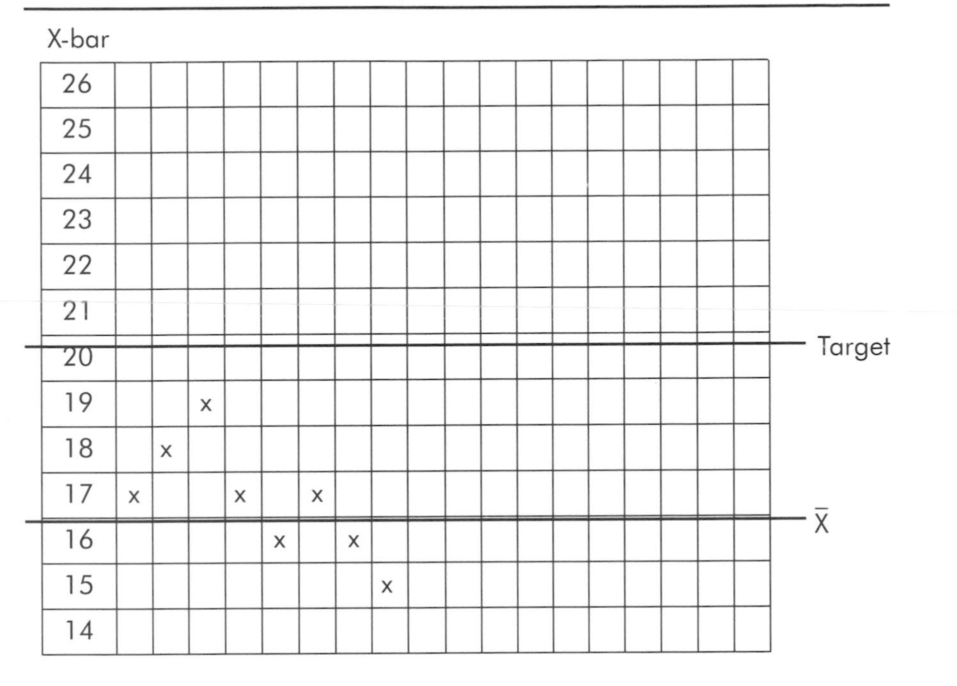

Range																		
8																		
7																		
6																		
5																		
4						x												
3	x		x	x			x											
2		x			x													
1																		
0																		

One of the next steps in developing a meaningful X-bar and R chart is to overlay customer specifications and upper and lower control limits onto the chart. This allows the relative position of what the customer wants against what the process is producing to be seen. If the process is trending toward one specification limit, or bouncing wildly across the target, then obviously some new measure of control needs to be established quickly. Otherwise, bad parts will be produced. Table 5-4 shows the customer-specification limits overlaid on the chart in Table 5-3. This tool is used to identify and record common causes (normal variation) or special causes (unusual variation) that could result in quality problems if not addressed. Machine wear, tool reliability, operator variation, and other factors can begin to show up as unacceptable variation or trends.

Software packages to help analyze trends are now available that contain some of the more complex statistical tools. Calculating sigma limits would be the next step here. The results are shown in Table 5-5.

Here, the sigma-limit calculator shows a sigma-limit value of 1.14 or a three-sigma value of 3.421. When these values are centered over the X-bar and overlayed onto Table 5-6, it can be seen that while there is relatively small variation within the three divisions of sigma, parts are falling far outside the acceptable limits established by the customer. This shows that the process is no longer capable. To satisfy the customer, the process must be brought back under control. Otherwise, bad parts will be shipped or they must be sorted out for repair or scrap. Neither is acceptable to an organization focusing on Six-Sigma performance.

The power of calculating a process sigma limit is that it allows you to see how closely to the target (or acceptable limits) the process is performing. With a little practice, you can begin to visualize how these limits would look when overlaid on a histogram, such as the one shown in Figure 5-4.

The principle of using sigma limits to visualize and control a process is closely related to the use of control limits (discussed earlier). The calculation for control limits can be contrasted with the sigma-limit calculations. There are many variations for calculating control limits. The following example is a variation that demonstrates the calculation for control limits for individual measurements (as opposed to subgroups of data).

Table 5-4. Simplified X-bar and R chart

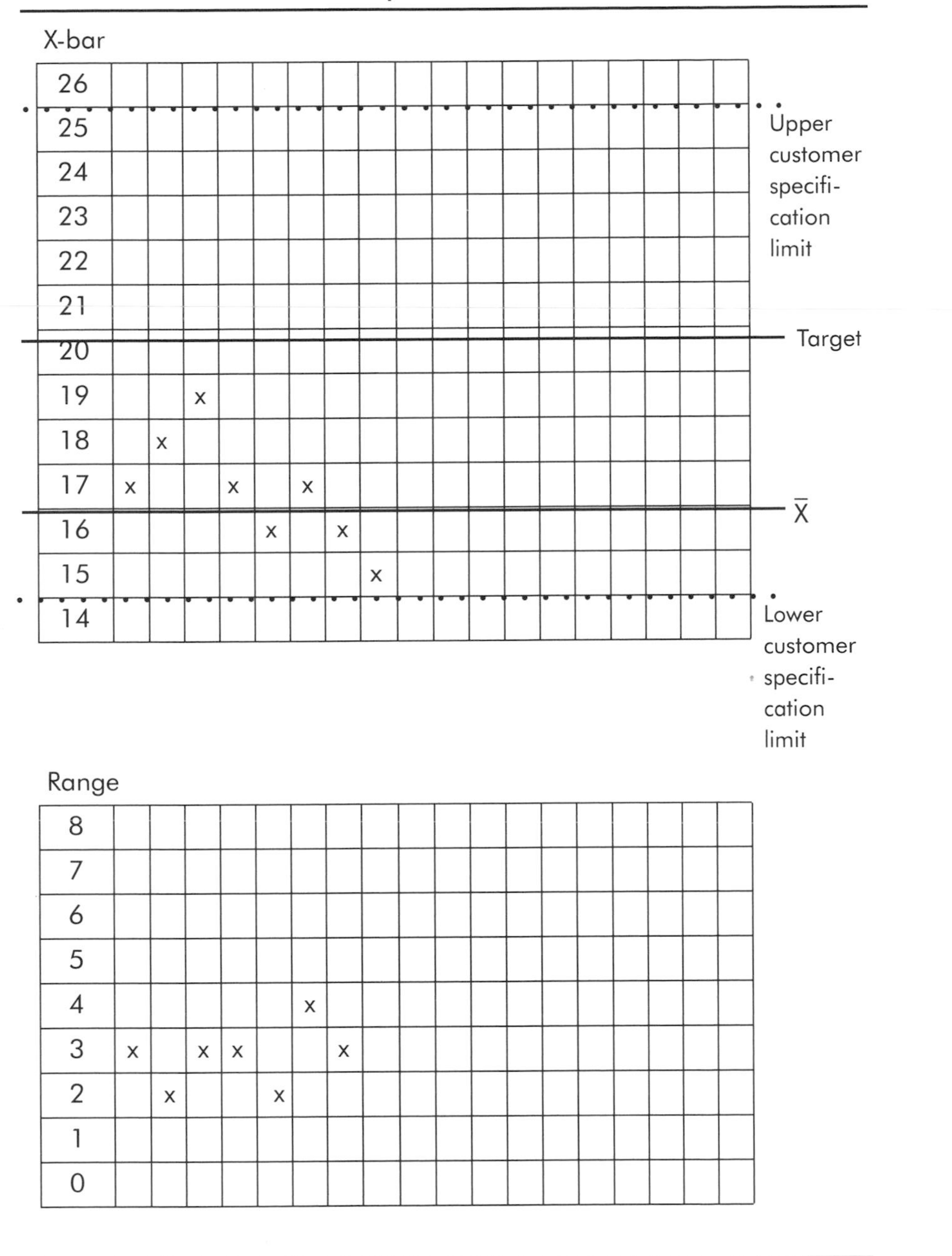

Table 5-5. Sigma limits calculator

Target	20.000
Upper tolerance	5.000
Lower tolerance	5.000
Upper specification limit (USL)	25.000
Lower specification limit (LSL)	15.000
Average	17.600
Sigma limit	1.140
3 Sigma	3.421
High	19.000
Low	16.000
Range	3.000
C_{pk}	0.760
C_p	1.462

For example, 25 cement blocks are weighed and result in the following values:

Weight: 18, 17, 16, 18, 18, 19, 21, 20, 21, 22, 20, 17, 18, 18, 20, 20, 21, 19, 19, 20, 18, 22, 19, 19, and 21 lb

Range: 0, 1, 1, 2, 0, 1, 2, 1, 1, 1, 2, 3, 1, 0, 2, 0, 1, 2, 0, 1, 2, 4, 3, 0, 2

The total (sum) of these values is 481 and the largest range is 6 (from 16 to 22).

To calculate the upper and lower control limits, first find the *average weight range*:

$$R = R_S \div R_N = 33 \div 24 = 1.375 \quad (5\text{-}2)$$

where:

R = average weight range
R_S = sum of ranges
R_N = number of ranges

Now, find the *average weight*:

$$\bar{X} = M_S \div M_N = 481 \div 25 = 19.24 \quad (5\text{-}3)$$

Table 5-6. Simplified X-bar and R chart

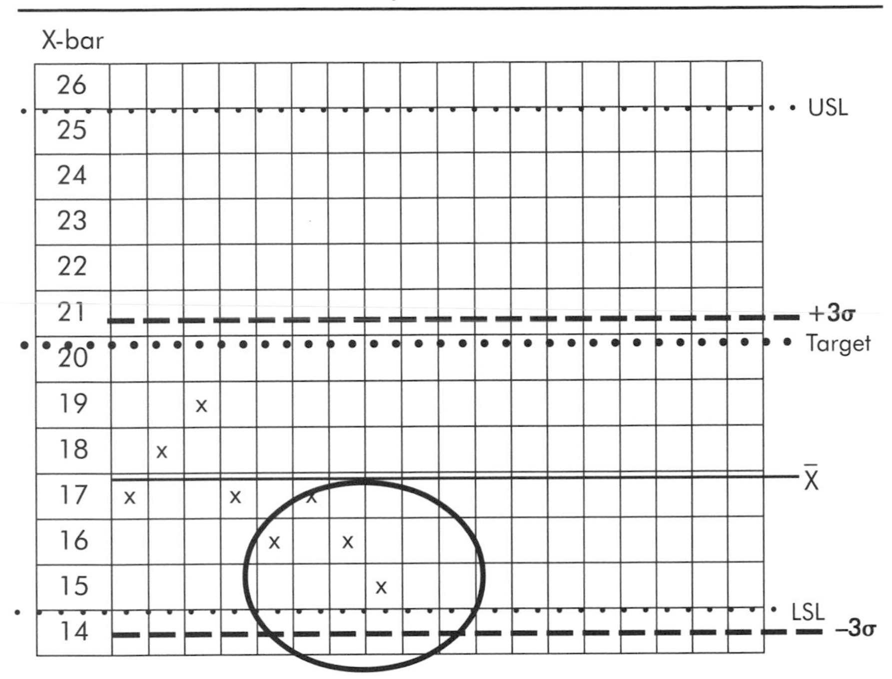

Range

8																		
7																		
6																		
5																		
4						x												
3	x		x	x			x											
2		x			x													
1																		
0																		

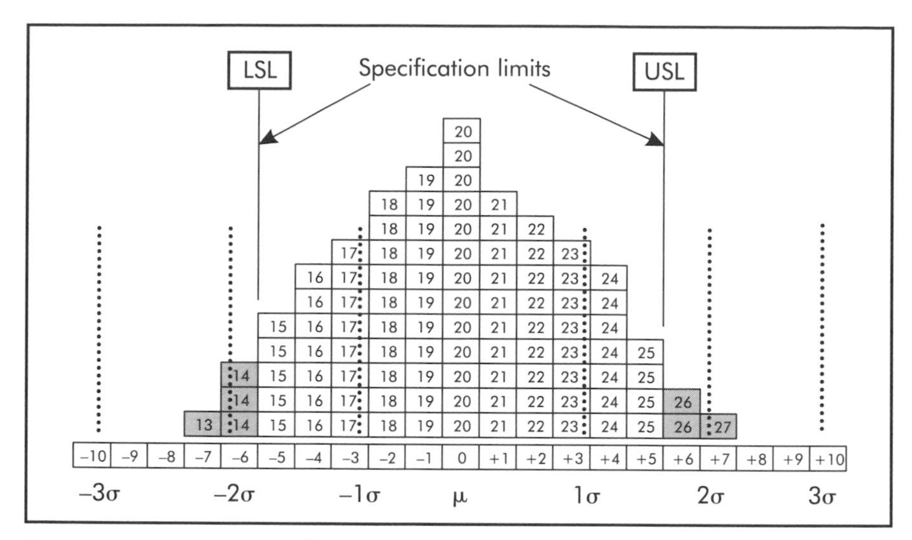

Figure 5-4. Sigma limits overlaid on histogram.

where:

$\bar{X}$ = average weight
M_S = sum of measurements
M_N = number of measurements

The formula for calculating the lower control limit (LCL) is:

$$LCL = \bar{X} - E_2R \quad (5\text{-}4)$$

where:

E_2 = constant from a statistical table = 2.66 (for individual data)

If the weight data is the average of a subgroup instead of individual piece weights, refer to a statistical table to find the appropriate constant. Such tables can be found in statistical textbooks.

Continuing the calculation:

$$LCL = 19.24 - (2.66 \times 1.375) = 15.583$$

The formula for calculating the upper control limit (UCL) is:

$$UCL = \bar{X} + E_2R \quad (5\text{-}5)$$

or

$$UCL = 19.24 + (2.66 \times 1.375) = 22.898$$

The upper and lower control limits are placed in their relative position to the target of 20. In this case, any part that weighs more than 22.898 or less than 15.583 is cause for concern.

A word of caution here. Some companies end up over-controlling a process (like over-steering a car on ice) when a sudden change is seen in one direction or the other on an X-bar chart. Some variation is to be expected, and if you start chasing normal process variation, you will end up over-controlling. Proper use of the X-bar technique teaches you to ignore normal variation and adjust the process only when a clear trend is sustained or when a special cause is identified. That is, a new batch of material or new operator performing in a way that produces a different result.

SPC for Small Lots of Parts

Make-to-order shops and job shops seldom run enough parts to justify an X-bar and R chart. Most of their jobs fall below the 25-unit lot size seen as a reasonable and statistically sound sample size. However, there is a way to track the variation of a few key (control) dimensions on parts over time to see how closely the process is running, no matter how many different parts are run. For example, let's say that three different part numbers were run today on the same machine. The first part has a key dimension of 4.125 that must be held to ±0.010, the second part a dimension of 9.375 held to the same tolerance, and the third a 0.667 dimension also to the same tolerance.

Each part can be tracked on the same control chart. After setting up the machine and verifying the part as acceptable, the operator measures the first five parts. If all five are acceptable, then the operator follows the normal inspection frequency required by the sampling plan outlined by the customer or the quality-assurance department.

The operator records the key dimension(s) on a common chart. A standardized sampling approach may require the operator to measure and record the first five parts and then every fifth part until 25 parts are run, then every 25th part (if there are that many parts being produced). When the operator moves from the set-up on part #1 to the new set-up on part #2, the process begins again, using the same control chart. The target for part #1 is different than for part #2, but the operator simply records the variation from the target dimension as shown in Table 5-7.

Control limits for small lots of parts can be calculated by dividing the customer specification into "fourths" and overlaying those values onto the chart as shown in Table 5-8. Since all the parts in the example have a tolerance of ±0.010, the tolerance is divided in half and overlayed on each side of the target. This determines a green zone (where everything is OK) and a yellow zone (where something is wrong). Use of this type of tool is probably the easiest and quickest to train plant wide.

If two consecutive parts measure in the yellow zone (same side), then an adjustment is needed. If any part falls outside the yellow zone (into the red zone), then an adjustment must be made unless it can be identified as a special cause (for example, an operator mishandled the piece). If any two consecutive pieces fall in the

Table 5-7. Short-run (multi-part) control chart

~														
+0.004														
+0.003														
+0.002	x		x											
+0.001		x							x					
Target	x			x		x		x		x	x	x		
–0.001					x		x						x	
–0.002														
–0.003														
–0.004														
~														
	Part #1	Part #1	Part #1	Part #1	Part #1	Part #1	Part #2	Part #2	Part #2	Part #3	Part #3	Part #3	Part #3	

Table 5-8. Short-run control chart with control limits

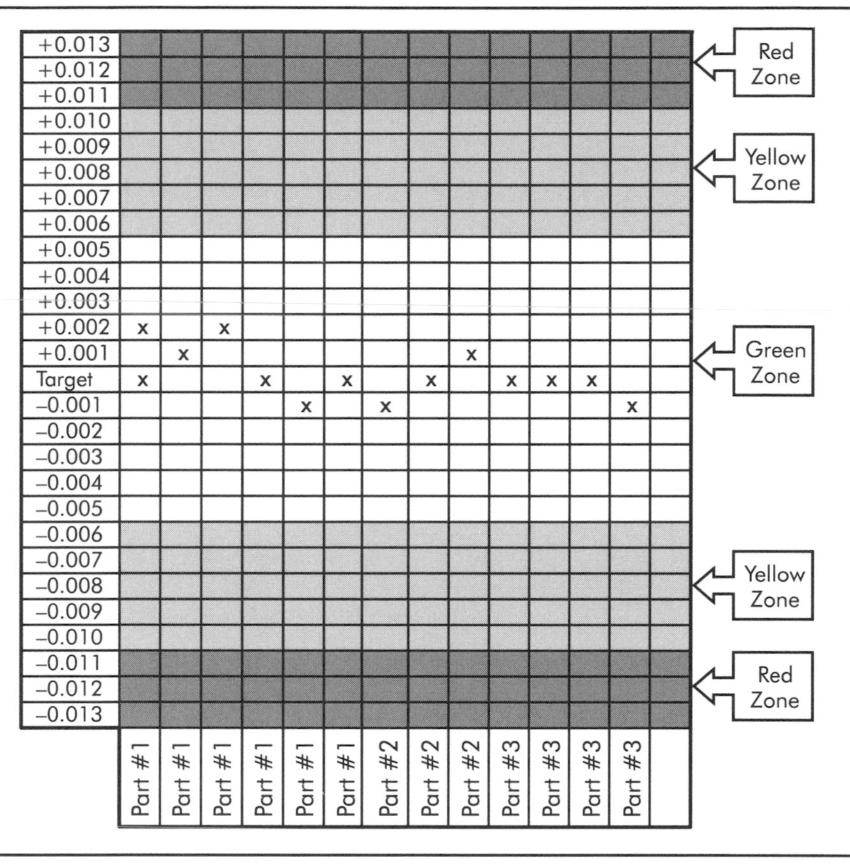

	Part #1	Part #1	Part #1	Part #1	Part #1	Part #1	Part #2	Part #2	Part #2	Part #3	Part #3	Part #3	Part #3	
+0.013														
+0.012														
+0.011														
+0.010														
+0.009														
+0.008														
+0.007														
+0.006														
+0.005														
+0.004														
+0.003														
+0.002	x		x											
+0.001		x							x					
Target	x			x		x		x		x	x	x		
–0.001					x		x						x	
–0.002														
–0.003														
–0.004														
–0.005														
–0.006														
–0.007														
–0.008														
–0.009														
–0.010														
–0.011														
–0.012														
–0.013														

yellow zone but on opposite sides, then there is a high degree of opportunity to generate a defect. Special attention must be given to future parts to ensure that no parts fall into the red zone.

Summary

Most Six-Sigma techniques are based on tools similar to those just explained. If you can add, subtract, multiply, and divide and calculate an average, you can use the majority of the tools in the SPC toolbox. Table 5-9 summarizes SPC's impact on Six Sigma. There are other SPC tools, but those just discussed are the most basic and the ones generally used by machine operators.

Table 5-9. Statistical Process Control summary

Relationship to Six Sigma	Fundamental to the theory of improvement through reduction of unwanted variation
Who needs and uses it	Managers, black belts, green belts, team facilitators, process owners, operators
Cost	Low to moderate, depending on the level of training needed
Strengths	Quantifies what is otherwise hard to measure or see in a process
Limitations	Does not fix anything—may not identify causes
Process complexity	Moderate (depending on experience or training)
Implementation time	1–6 months
Additional resources	See Bibliography
Internet search key words	SPC, Statistical Process Control, Statistics
Internet URLs	www.margaret.net www.minitab.com www.statsdirect.com

SPAGHETTI DIAGRAMS

The distance material travels in the manufacturing process is critical to the measurement phase. It is one of the easiest measurements to obtain, yet it is often ignored. There is a decades-old Mercury Marine video demonstrating the beginning stages of its improvement effort. The "before" condition shows one component of an outboard motor traveling nearly 4 miles (6.4 km) through the plant in 122 steps, only 27 of which were value added. When this example was multiplied by the hundreds of parts in an outboard motor (or the number of component parts the company processed in a month), it was no surprise things cost so much. Mercury obviously addressed the reasons for so much travel and made major improvements, but until they mapped it out, the opportunity was hidden from sight or solution.

The spaghetti diagram (Figure 5-5) is a tool employed by Kaizen and set-up reduction teams to show the distance traveled by op-

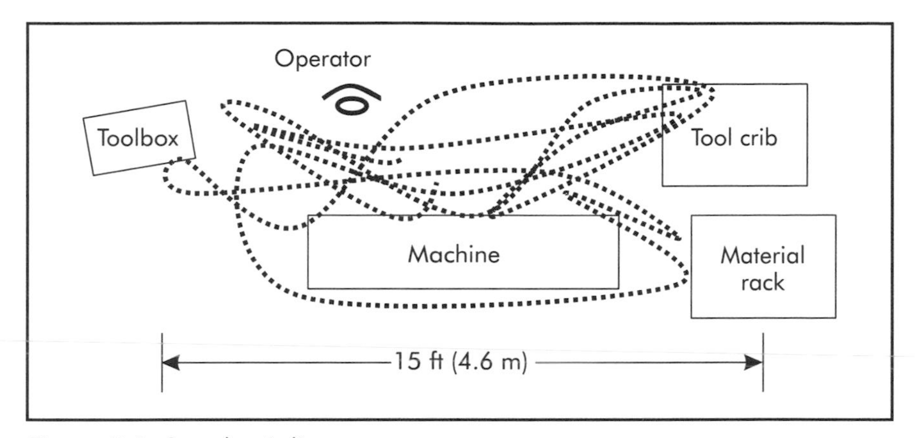

Figure 5-5. Spaghetti diagram.

erators while gathering tools, materials, and paperwork during the changeover from one part to another. They are also employed when measuring the distance traveled by parts in the manufacturing process. Teams have also used this tool when creating a visual representation of paperwork flow through an office or engineering department. It is an eye-opening experience to recognize how many times a document is passed from person to person or from desk to desk.

A spaghetti diagram is a great tool because anyone can do it. All that is needed is a simple sketch of the work area or floor plan and a pencil. The pencil is moved every time the operator or material moves (whichever one is being followed), then the total distance traveled is added up. The results are amazing. To put this into perspective, consider how far Bob travels in a year's time just setting up his machine tool. If he travels 150 ft (45.7 m) whenever he performs a set-up, performs six set-ups per day, and if he works 250 days per year, then: 150 × 6 × 250 = 225,000 ft (68,580 m) per year or nearly 40 miles (64.4 km). Think how long it takes to walk 40 miles (64.4 km). Even at a brisk 4 miles/hour (6.4 km/hour), that is 10 hours of Bob's work year spent just walking.

Working with a Kaizen team in a small casting firm, it was calculated that an average-size lot of parts moved an average of 1,680 ft (512 m) through the plant. The team moved an average of 72 lots per day, and worked 250 days per year. Therefore, those parts traveled 30,240,000 ft (48,656,160 m) or 5,727 miles (9,215 km).

Even though technology has allowed the elimination of some of the paperwork, the path taken by information can also be inordinately long. The spaghetti diagramming process can make visual what would otherwise go unnoticed. It brings an element of measurement (hard data) to non-value-added activities that too often are taken for granted or accepted as a given. The spaghetti diagram tool is summarized in Table 5-10.

Table 5-10. Spaghetti diagram summary

Relationship to Six Sigma	Helps measure distance traveled for operator, material, or paperwork
Who needs and uses it	Value-stream managers, process owners, black belts, green belts, improvement teams
Cost	Low
Strengths	Great visualization tool
Limitations	Does not change anything on its own
Process complexity	Easy
Implementation time	1 day
Additional resources	See Bibliography
Internet search key words	Spaghetti diagram
Internet URL	phe.rockefeller.edu/ie_agenda/ie2.html

VALUE STREAM MAPPING

Simply put, Value Stream Mapping (or value chain mapping) makes visible what would otherwise be invisible. *Lean Manufacturing* is a systematic approach to the identification and elimination of waste in all its forms. Value Stream Mapping helps to identify the waste—not just process waste, but waste associated with transportation, information flow, imbalances in work assignments, waiting, storage, and paperwork.

Through a series of symbols, lines, and text, the value-stream map allows you to see where inventory resides. By cutting inventory levels in half, manufacturing lead times can be cut in half (at least theoretically).

The value-stream map can be relatively simple as shown in Figure 5-6, or very complex as shown in Figure 5-7. These two current-state maps are generic examples of the complexities to expect within even relatively simple processes. The Value Stream Mapping process allows teams to meet the first rule of Kaizen, "know the process."

A value-stream map also allows process owners to examine in graphic form where inventory resides. Dividing inventory by operation process time, as in Figure 5-8, shows how many days worth of excess inventory is in the process, and helps calculate the amount that should be there. In Figure 5-8 there are 40 lots ahead of process #1 and 25 lots ahead of process #2. Because the first operation only takes five minutes to process a unit, and with an average of 50 units in a lot, one lot is produced every 260 minutes (4.33 hours). Then, why is there 10,400 minutes (173 hours) worth of inventory ahead of this work center? Answer: no one considered it important enough.

In Figure 5-9, the inventory for both work centers has been reduced to just one lot for the first operation, which only works one shift, and two lots ahead of the second operation, where two shifts are required due to the longer process time. This has the effect of taking 370 hours of lead time away (15 days). If each unit produced has an average value of $50, then reducing the inventory by 3,100 units (62 lots of 50) is equal to freeing up $155,000 currently tied up in inventory. It also saves $23,250 in interest and carrying costs if these equal 15%.

The Value Stream Mapping tool is summarized in Table 5-11. All companies need to take the step of documenting the current and future states.

THE DASHBOARD

When is the last time you ran out of gas in your car? I would guess everyone has had the experience once or twice. As a result, you are much more aware of the gas gage from that point forward (at least for a while.) I once had a 1970 Ford pickup that didn't have any dashboard lights, and it drove my wife crazy. It didn't bother me, but she couldn't understand how I could drive without a speedometer. She had a point. Measuring is very important when you

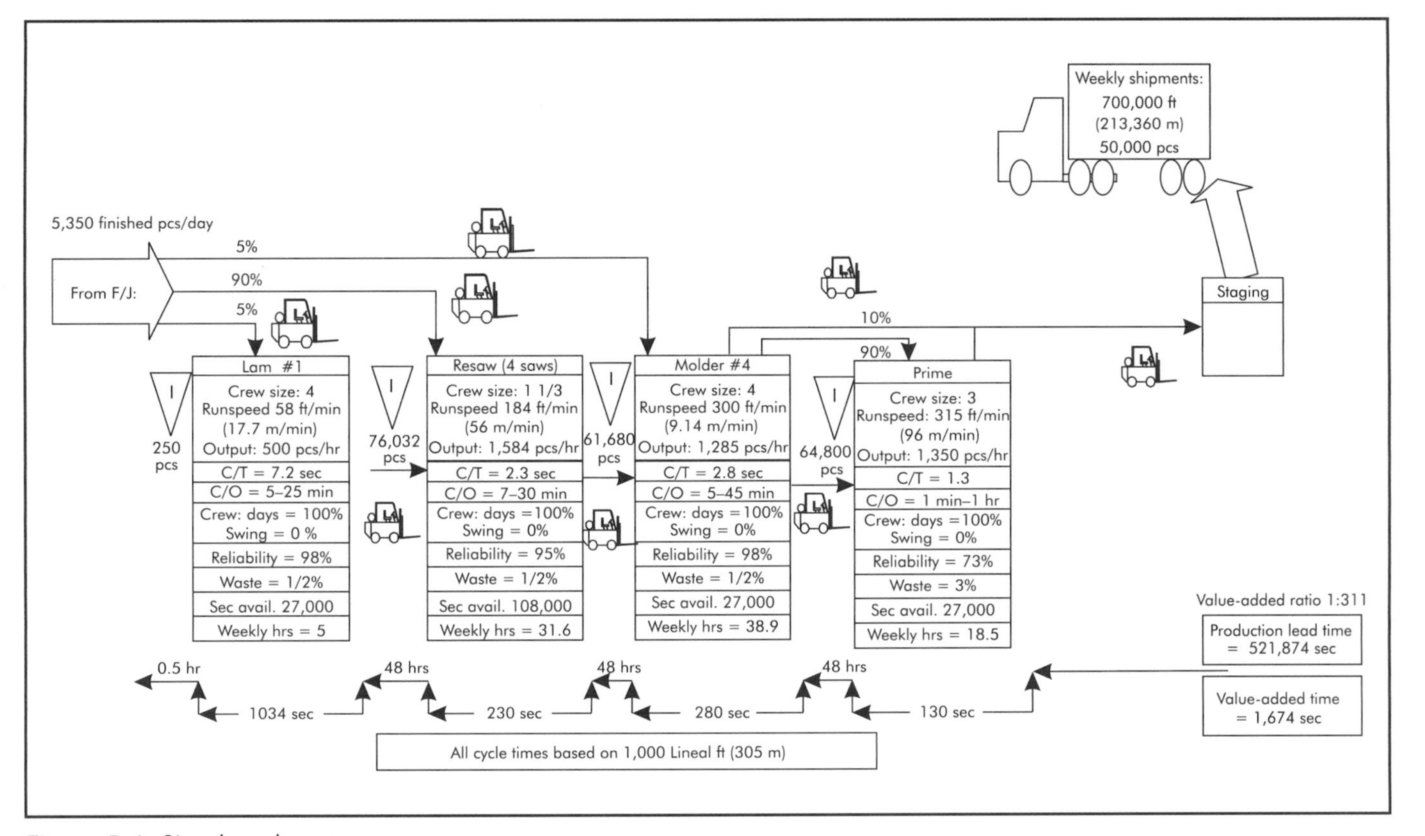

Figure 5-6. Simple value-stream map.

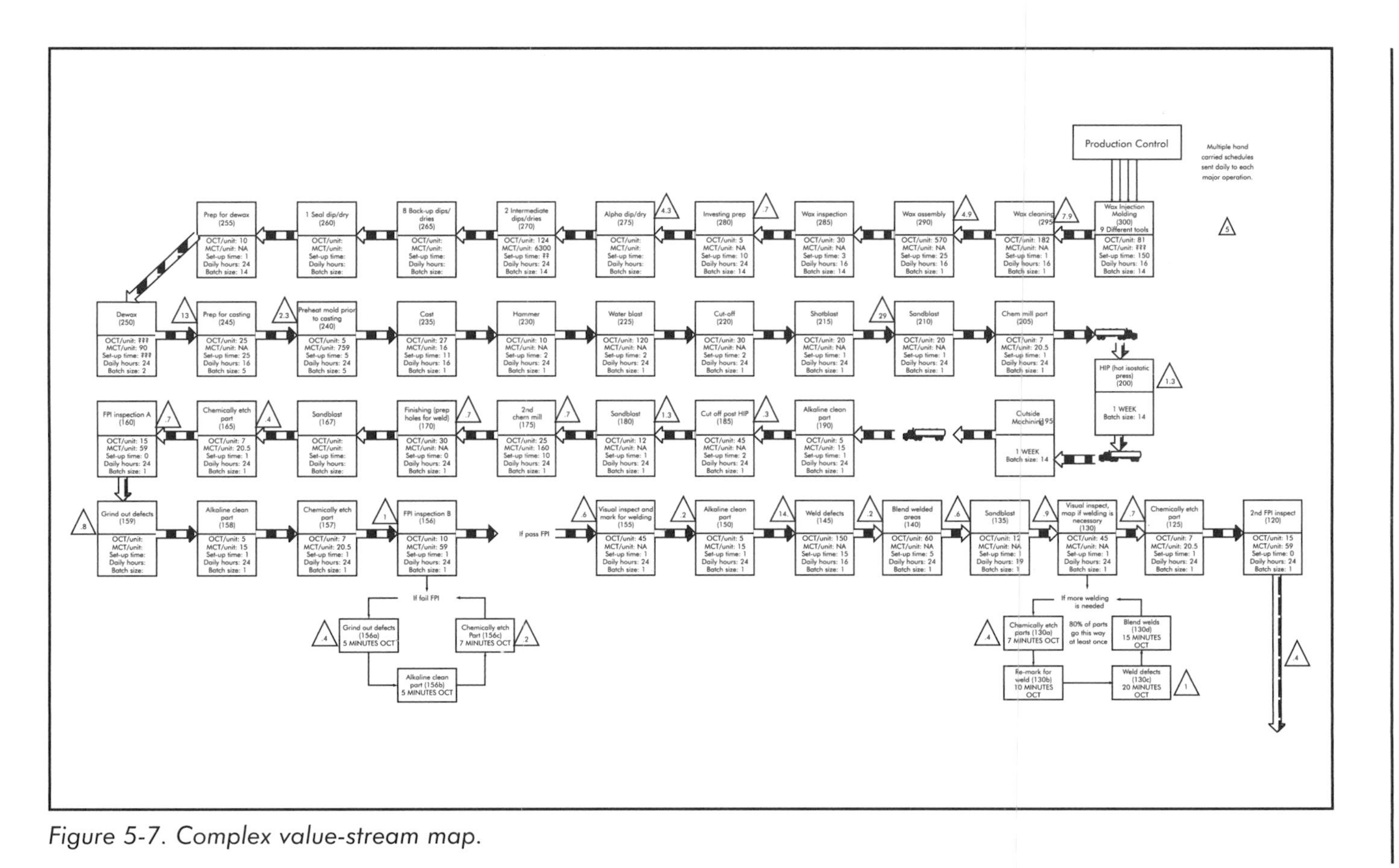

Figure 5-7. Complex value-stream map.

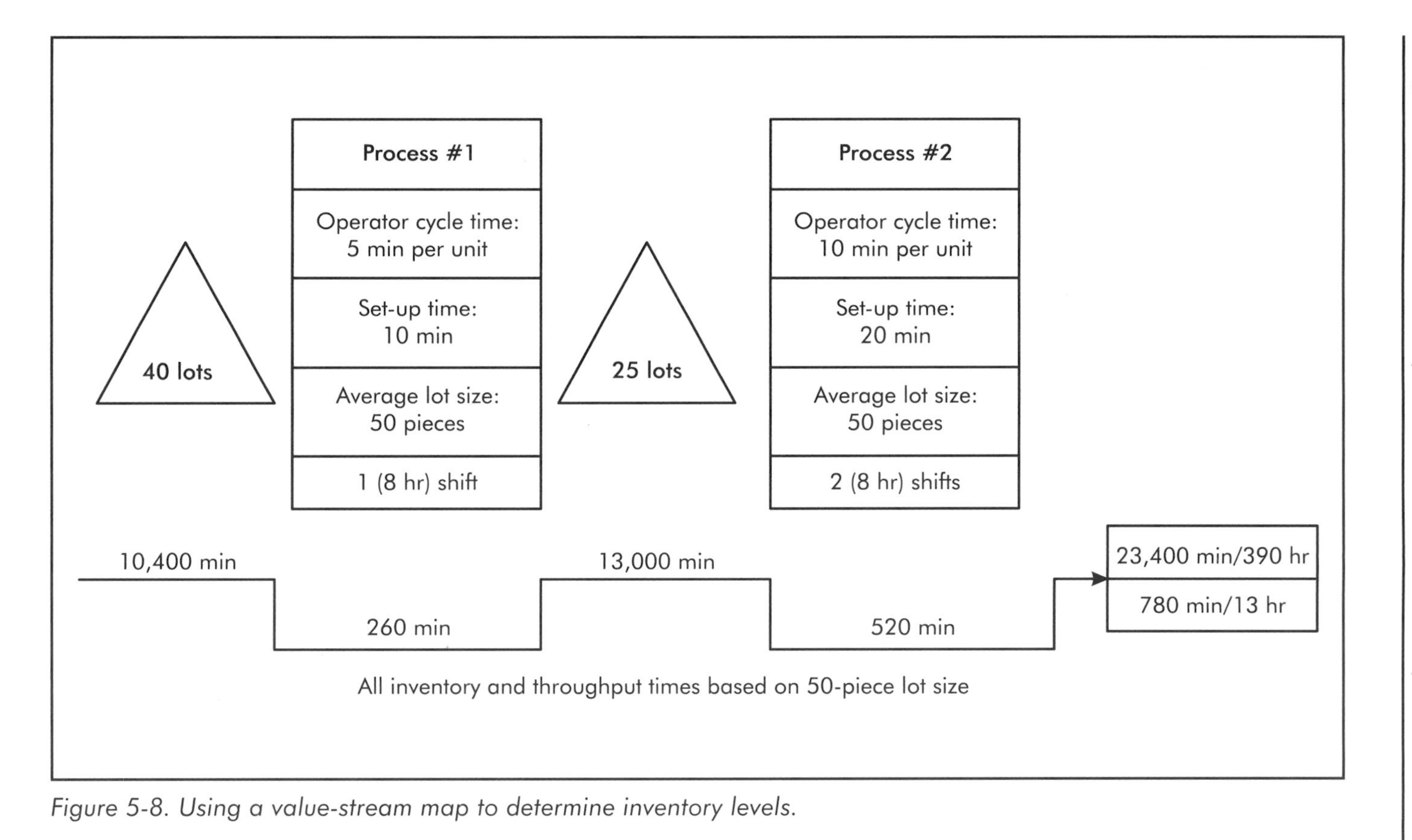

Figure 5-8. Using a value-stream map to determine inventory levels.

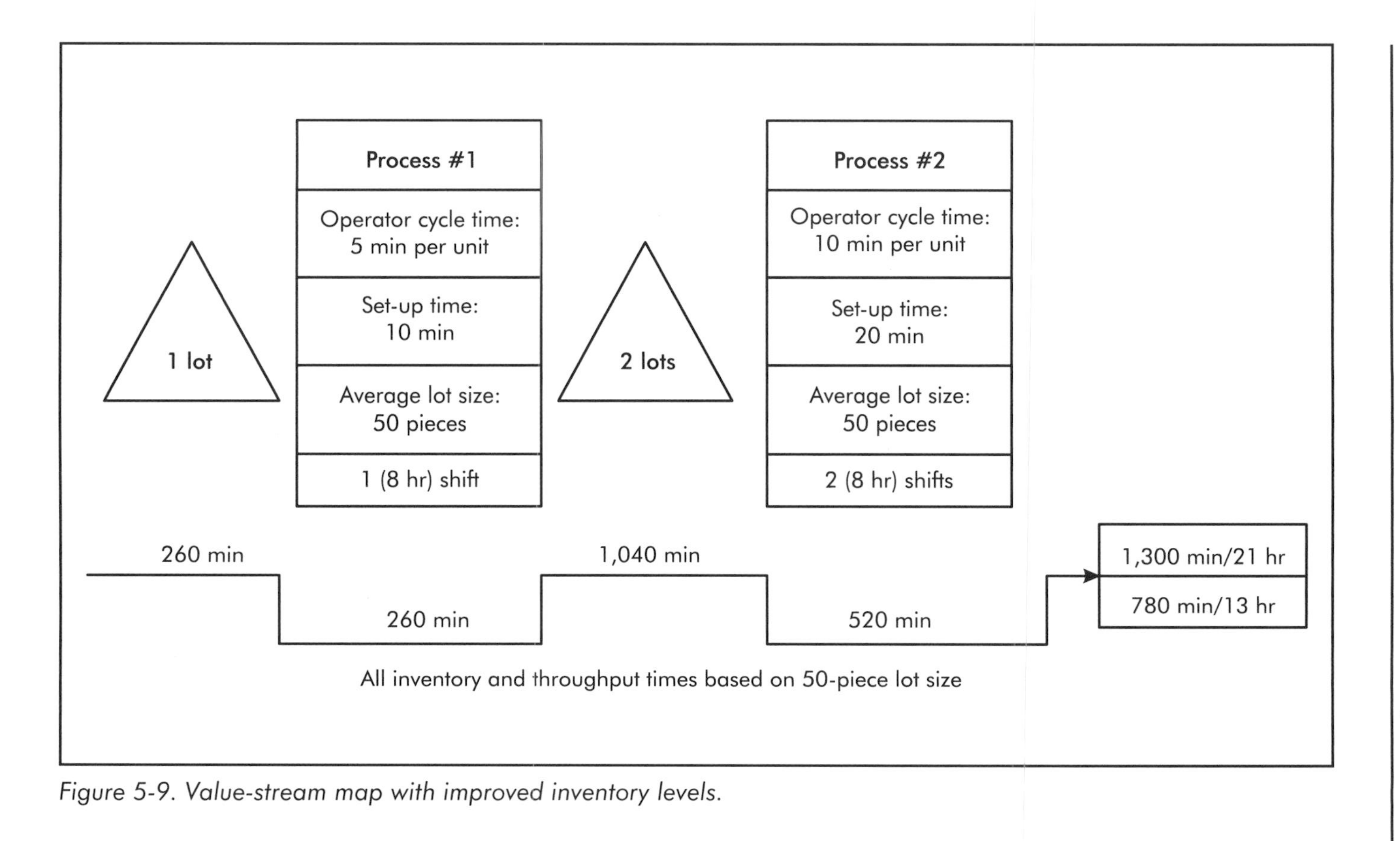

Figure 5-9. Value-stream map with improved inventory levels.

Table 5-11. Value Stream Mapping summary

Relationship to Six Sigma	Value Stream Mapping permits evaluation of each product or service stream and sets priorities during the Hoshin Kanri process. It defines how far a company has come.
Who needs and uses it	Steering teams, value-stream leaders, black belts, Kaizen teams
Cost	Low
Strengths	Very powerful tool for making inventory visible and painful
Limitations	Does not usually equate a dollar figure to the inventory, yet is an excellent tool for showing people how much money is sitting around
Process complexity	Medium
Implementation time	1–4 weeks
Additional resources	See Bibliography
Internet search key words	Value Stream Mapping, VSM, learning to see
Internet URLs	www.lean.org phe.rockefeller.edu/ie_agenda/ie2.html www.qualitydigest.com/march97/html/f3.htm

are hurtling down the road, and it is equally important to be measuring business performance.

Measuring what? How often should you measure? Who should measure it, and who should review it? There are week-long seminars and entire books written on the subject of finding the ideal performance metrics to track. Of course, you want your business to run like a well-oiled machine (as well as your car). A method of measurement that is visual, easy to interpret, and warns you immediately if something goes wrong is a means toward that end.

In a car, there are a few key gages: fuel level, water temperature, oil pressure, and speedometer. They tell a lot about a car's performance. As depicted in Figure 5-10, the common dashboard

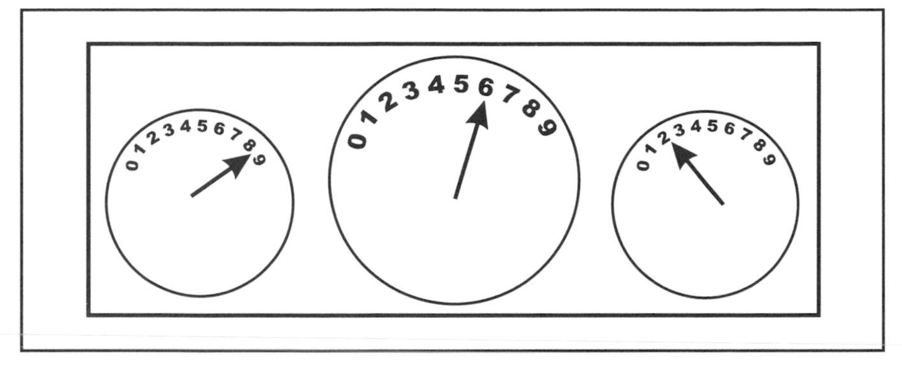

Figure 5-10. The dashboard.

tells at a glance how everything is doing. Only when these readings seem out of whack do you pay much attention. Interestingly, racecar drivers have little time to digest gage readings, so they turn them so that with a normal reading the gage is pointing straight up (Figure 5-11). This allows them to focus on the race in front of them.

The same applies to business. Companies don't win in business by spending a lot of time developing and reading charts. The right charts must be used to make good decisions and help steer the business. The process must not be overdone. If the dashboard in your car looked like the cockpit of an airplane with hundreds of gages, you would be less able to focus on your driving. So too must the measurements at work be manageable and not overwhelming.

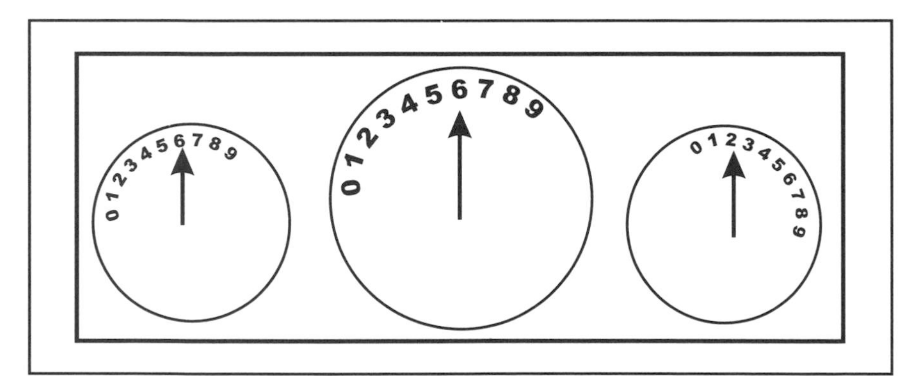

Figure 5-11. Improved racecar dashboard.

There are different levels of risk associated with different business variables, so the choice to measure something or not should be based on how important it is to the health or future of the company. Teams cannot spend all their time collecting data and measuring things just for the sake of measuring. They must expend their time improving a few "key" measurements.

How do you know if the company is measuring the right thing? Things do not improve simply because they are measured. A measurement should relate to something that can vary and that can be controlled. It should give notice when something is awry and allow the person responsible for that variable to make a correction to the process if a measurement falls outside a predetermined range of acceptability.

Companies taking the high road to World-Class performance have stopped focusing on measurements like machine up-time and instead are measuring how often the machine is making the right part at the right time (neither too late or too early). Departmental measurements where only labor and overhead are measured have gone by the wayside in favor of measuring overall effectiveness.

In a search for sterling quality, the new measurement is related to first-pass or first-time-through capability. In an effort to key in on velocity, World-Class teams track dock-to-dock time and the total cost to manufacture a product.

The CEO will no doubt have a different set of gages on his/her dashboard than the vice president of operations. And the value-stream leader's dashboard also will be different. The key is to measure a few things (the right things) well. Here are some examples of items that can be measured:

- growth in earnings;
- return on equity, capital, and assets;
- manpower required to produce product;
- customer satisfaction;
- reorder rates;
- materials required to produce product;
- work-order accuracy (other documents required);
- supplier lead times;
- supplier quality;
- productivity per team member;

- inventory turns;
- on-time delivery performance;
- energy costs;
- warranty costs;
- employee satisfaction;
- engineered drawing accuracy;
- process cycle time;
- process waste;
- process quality (defects per million operations);
- variance in machine performance;
- training quality/quantity; and
- cross-training/cross-functionality.

There are dozens more, but the point is that if a company spends a lot of time collecting all this data, there could be no time left to improve anything. A better approach might be to use the same method of monitoring the business that is used to monitor an automobile's performance. A few essential gages are all that is needed. When the key metric (dashboard needle) jumps or drifts toward the upper or lower limit of acceptability, the same kind of techniques can be used as those that might be applied if a car's temperature gage started reading too hot.

With a car, troubleshooting techniques are applied to narrow down the problem. These tools might include checking water levels, hose leaks, belt tension, thermostat, water pump, etc. The entire process would be audited until the cause was found. Why should it be any different in business? Why should a company spend a lot of time collecting data unless there is an indication of a potential problem? The trick is to measure the right thing. In the car example, gages such as a belt-tension measurement device, a pressure gage on the water pump, and level gages for the radiator could be added. Yet, what's the point? Why spend time installing and monitoring all these new gages when all that is needed is the temperature gage, which is there already? It will signal when there is a cooling-system problem, and then the necessary detective work can be done to find the problem.

Finding the one or two gages that signal an impending problem is a better solution. Even having the right gage, however, will fail to flag a problem if it is ignored. It is only of value if it is checked on a routine basis. Then, like a car's temperature gage, once it points out a potential problem, a company can act fast before damage is done.

Most companies have their middle managers spend inordinate amounts of time generating reams of paper to qualify what they are doing for the benefit of upper management. Why not allow these managers to spend their time helping the people closest to the customer and his or her needs: the operators of the process. Find a measurement that tells how often the desired output happens and monitor it.

The dashboard development process consists of five phases:

1. Design: Focus on what, how, and who. Develop strategies, goals, and reports. Identify the sources of information. Define priorities, success factors, and measurable goals. Define the implementation approach.
2. Implement: Develop the conduit of information. Information rolls up from the customer through the manufacturing process, then through production management, engineering, order entry, and sales. Each person in this chain of information has a slightly different dashboard that applies to those business elements within their control.
3. Test drive: Ensure that the dashboard indicates the key metrics associated with the company's vision. Check that it is easy to use and cost effective.
4. Run: Fully integrate the program throughout the organization. This guarantees ownership of the goals and accountability for results.
5. Evaluate/recalibrate: Every quarter, all dashboards should go through a quick check-up. Evaluate the fit between strategies, information relevancy, and feedback value.

Figure 5-12 shows some examples of dashboards. In summary, make sure the measurements are simple, based on something that is controllable, and measure the most important things. Identify the variables that tell the most about where to dig (in the form of an audit) if problems surface. The dashboard tool is summarized in Table 5-12.

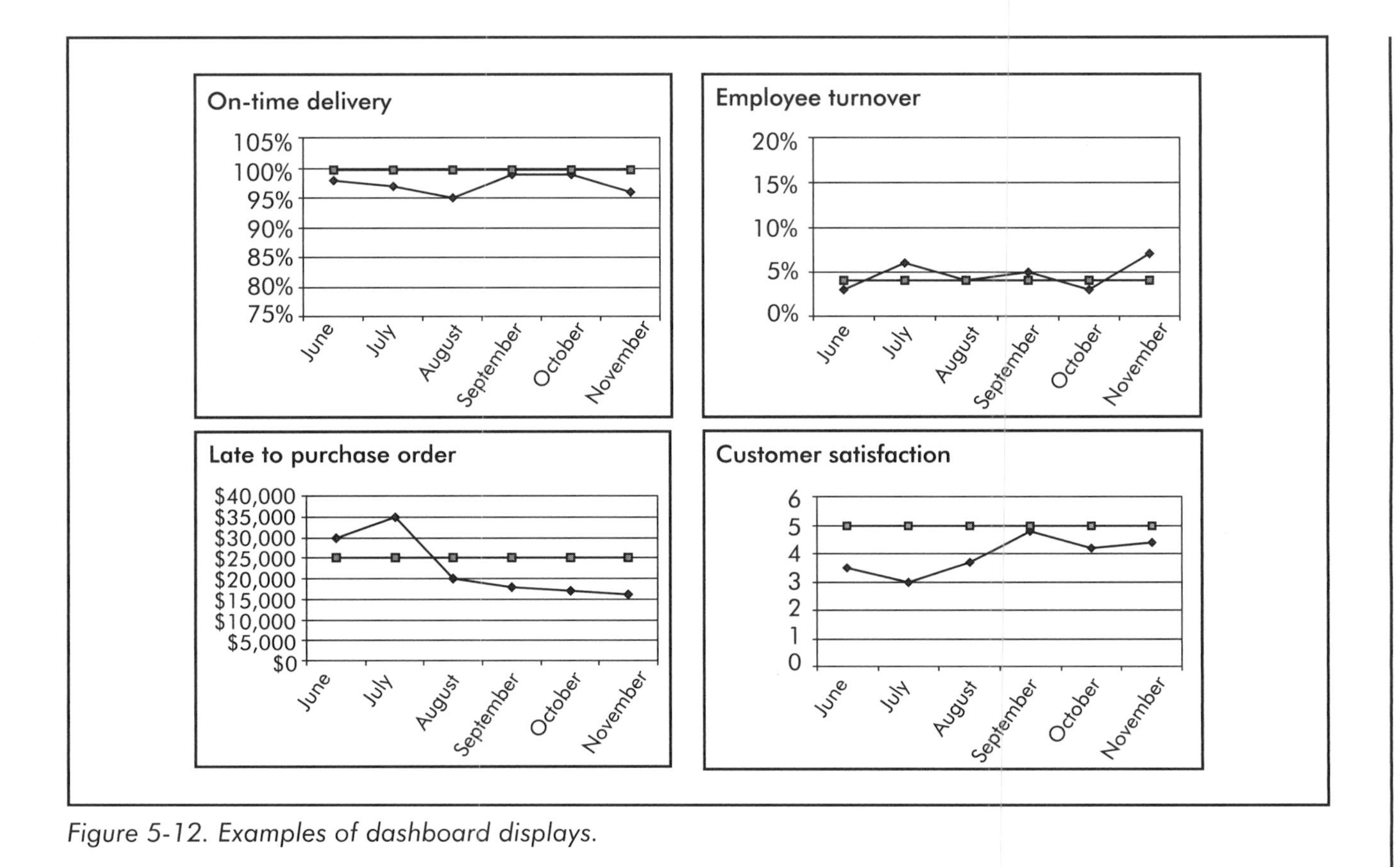

Figure 5-12. Examples of dashboard displays.

Table 5-12. The dashboard summary

Relationship to Six Sigma	An effective approach to measurement, a key aspect of Six Sigma
Who needs and uses it	Executive management, value-stream leaders, department heads
Cost	Moderate (due to programming costs)
Strengths	Helps to quickly identify unexpected changes
Limitations	Only helps identify problems and does not provide troubleshooting and root-cause analysis
Process complexity	Medium
Implementation time	1–3 months
Additional resources	See Bibliography
Internet search key words	Executive dashboard
Internet URLs	www.hpctech.com/auxilor.nsf www.emeraldhillsstrategy.com/dashboard.html www.lingocd.com/exec.htm www.shopwerkssoftware.com/pip.htm

CONCLUSION

At the end of the Measure step, the deliverables should include the definition of the problems that are potential causes of defects, and identification of the variables or conditions that could lead to such defects.

There should be an identification, verification, and prioritization of input, process, and output measurements that indicate when a defect is generated. A recommendation for methods of monitoring important elements in the process is another deliverable from this step.

Prior to moving to the Analysis step, the measurements should be distilled down and organized.

Suggestions or examples for how to measure, collect, and collate data on an ongoing basis should also be an outcome of the Measurement step.

6

Step Three: Analyze

The purpose of this chapter is to point out tools that will help distill all the measurements taken and begin the analysis process. The type of tools range from simple ones like asking elementary questions to far more complex analysis techniques. It is surprising how often the simple, yet powerful tools are overlooked. Thousands of small problems can be solved and root problems identified by using what should be intuitive techniques. Too often things are made more complicated than needed.

To make the right decision, data from the process must be evaluated. Design of Experiments can demonstrate how to identify variables within the process, and how changing them can produce different controllable results. The Taguchi method provides engineers with the same tools in the design phase as in the manufacturing phase and, thus, can eliminate the chance for variability even before the job hits the manufacturing floor. Closely related to concurrent engineering is the Theory of Inventive Problem Solving (TRIZ). Finally, the use of quality mapping, a close cousin to Value Stream Mapping, rounds out the discussion of analytical tools. By no means does this chapter represent every analysis tool available.

The goal of performing at Six-Sigma levels is unattainable if a company is not routinely analyzing the entire value stream and identifying new opportunities for improvement. A meaningful Six-Sigma program cannot be based on the idea of sweeping through the plant one time, and then considering the task complete.

THE FIVE WHYS (PLUS HOW)

My four-year-old son knows the power of the Five Whys technique, and uses it often. The inquisitive nature of children is something that we would do well to imitate when attempting to find root causes and solutions to problems. As adults, people often fail to challenge a process or simply ask "Why?"

The Five Whys technique is a powerful means of identifying the true root cause of a problem as well as gaining a description of the process itself. By asking open-ended questions and then following up with another open-ended question (one that requires more than just a yes or no response), the root of the problem is usually discovered.

For example:

- "Why are these parts in the recycling container?" Because the cement blocks vary in weight.
- "Why do the blocks vary in weight?" Because there are so many different operators.
- "Why should it matter that there are different operators?" Each operator can use his or her own recipe and process.
- "Why isn't there a standard procedure so everyone uses the same process?" No one has ever seen the need to develop one.
- "Why hasn't someone seen the need for a standard to be developed?" The company has not tracked this form of waste in the past because it was not considered an important issue.

A variation on the Five Whys is the Five Ws (Who? What? When? Where? Why?), and then the question "How?" is added. The *Five Whys* (or Ws) identify the problem and the *how* begins the analytical process. Very little will remain undiscovered if answers are provided to each of these questions for each variable in the process.

- What kinds of defects are possible here?
- When does that kind of defect happen?
- Why do you think that is?
- Who controls this particular variable: the machine or the operator?
- Where is the adjustment for that variable?
- How can we better control that adjustment?

The Five Whys process is summarized in Table 6-1.

DESIGN OF EXPERIMENTS

Design of Experiments (DoE) is an approach for exploring the cause-and-effect relationships between at least two variables.

Table 6-1. The Five Whys process summary

Relationship to Six Sigma	Great tool for seeking quick solutions to problematic issues
Who needs and uses it	Everybody
Cost	Low
Strengths	Quickly generates new ideas and encourages people to think beyond normal processes
Limitations	The Five Whys process can generate too many opinions and not enough facts. Opinions must be tested before action is recommended.
Process complexity	Easy
Implementation time	1 day
Additional resources	See Bibliography
Internet search key words	TQM tools, wwwwwh
Internet URL	members.aol.com/kaizensepg/qtools.htm

Through proper design, record keeping, and testing, interrelationships can be examined, priorities between variables can be established, and risk or loss can be accurately predicted from changes in essential process variables.

Although it is normally best when analyzing a process to vary only one input at a time to observe its effect on the rest, a designed experiment allows for controlled management of more than one variable during a test. This method works best when each variable has only two conditions, such as "on" and "off." However, the test examines the effects of all potential combinations of variables (inputs) under consideration.

Although the obvious benefit is that a single experiment takes less time to run (and ties up less valuable production time) than running individual experiments for each variable, it does add a level of complexity that must be acknowledged. This complexity requires any random factors not being measured to be controlled (held constant) or otherwise accounted for during the test to get valid results.

An important issue is the total number of combinations of variables to be observed during the experiment. For example, if there are three variables and each variable has two conditions, then, there are eight possible combinations that need to be tested.

For example, assume the fictional concrete-block manufacturing company wants to test the effect of a basic recipe for cement on the output of blocks. To test the strength of a concrete sample, various recipes of concrete are mixed, each is poured into a mold, and all the concrete samples are dried for exactly the same period. Then, the test pieces are subjected to a hydraulic press to measure the pressure (psi [Pa]) that each piece will bear before failure.

Table 6-2 shows the results for every possible combination of the three ingredient variables for two possible conditions: 1 lb (0.5 kg) or 1.5 lb (0.7 kg). This experiment is called a "3 × 2" because there are three variables and two possible conditions. A 2 × 2 would have only two variables (for example, changes in the water and concrete amounts only, with a constant amount of sand).

According to Table 6-2, test #2 yielded the strongest concrete. However, if the customer specification calls for only 1,500 psi (1.0 Pa) strength or better, there may be significant cost to adding more concrete than required to meet the specification. All such considerations need to be included.

Table 6-2. 3 × 2 (3 factors × 2 variables) design of experiment

Test #	Factor #1 Concrete Amount, lb (kg)	Factor #2 Water Amount, lb (kg)	Factor #3 Sand Amount, lb (kg)	Result Strength, psi (Pa)
1	1 (0.5)	1 (0.5)	1 (0.5)	1,800 (1.2)
2	1.5 (0.7)	1 (0.5)	1 (0.5)	2,200 (1.5)
3	1 (0.5)	1.5 (0.7)	1 (0.5)	1,600 (1.1)
4	1.5 (0.7)	1.5 (0.7)	1 (0.5)	1,500 (1.0)
5	1 (0.5)	1 (0.5)	1.5 (0.7)	1,200 (0.8)
6	1.5 (0.7)	1 (0.5)	1.5 (0.7)	1,400 (0.9)
7	1 (0.5)	1.5 (0.7)	1.5 (0.7)	1,600 (1.1)
8	1.5 (0.7)	1.5 (0.7)	1.5 (0.7)	1,800 (1.2)

Of course, there are other variables that could affect the quality and strength of the concrete: mixing time, ambient temperature, drying time, etc. The number of variables (factors), possible conditions, and the multiplying effect of each will determine the complexity of the experiment.

Replication of the experiment may be necessary over time to ensure that test results can be duplicated given the same factors or variables performed. This might include a test on the same equipment at a different time, by different operators, or on a similar set of machines by the same operators. Thus, it is a good idea to run a random test to measure or audit the assumptions made during an experiment to validate the conclusions.

The Taguchi Method

Developed by Dr. Genichi Taguchi, the Taguchi method refers to a set of techniques used in quality engineering. These tools embody both statistical process control (SPC) and many new quality-related management techniques.

Process costs and repeatability problems can cripple new programs right at the starting line. Poor design and poor process methodology can result in major financial losses over the life cycle of a product. Small incremental improvements in the manufacturing processes, while useful in the short term, may not provide the required levels of quality, reliability, or economy of production needed over the long term.

Taguchi methods attempt to ensure quality through design and the identification and control of critical variables that cause deviations to occur in the process, product, or service quality.

Most of the attention and discussion on Taguchi methods are focused on statistical tools and evaluated with the terms "robustness," "robust design," or "robust manufacturing." A robust process is one that is not easily upset or changed. Other synonyms for robust include vigorous, hardy, stalwart, strapping, sturdy, strong, or sound. The question to ask is: "Are the design and manufacturing processes sound, strong, and not easily overturned?" If so, they can be considered robust.

The Taguchi concept is based on these basic ideas:

- Quality is not measured by how close to the tolerance (band) a measurement is, but rather by how far it varies from a very specific target.
- Quality must be built into the design and the process. Quality can not be assured through inspection and rework.

Taguchi pointed out that SPC only eliminates faults and defects if they are detected, and only after they are manufactured. A better approach is to use a methodology to prevent the defect from occurring in the first place. Taguchi further argues that through a systematic approach to the design, processes can be made insensitive to variations (which he calls noise, based on his background in audio engineering). This avoids the costly results of rejection and rework. Design of Experiments and the Taguchi method are summarized in Table 6-3.

Table 6-3. Summary of Design of Experiments and Taguchi method

Relationship to Six Sigma	Higher-order statistical analysis tools for identifying and controlling key variables in a process
Who needs and uses it	Black belts, green belts, project teams
Cost	Moderate, depending on the complexity of the test
Strengths	Removes doubt about the true cause and effect (and interrelationship) of multiple variables
Limitations	Few weaknesses or limitations apply to either approach.
Process complexity	Moderate to difficult
Implementation time	1 week–6 months, depending on the complexity of the process being tested
Additional resources	See Bibliography
Internet search key words	DoE, Design of Experiments, Taguchi
Internet URLs	www.umetrics.com www.sigmazone.com www.sinc.sunysb.edu/Stu/mphagoo/bhe.html www.quality.org/Bookstore/Taguchi.htm

TRIZ (THEORY OF INVENTIVE PROBLEM SOLVING)

Yet another new wave and buzzword to learn about is TRIZ (pronounced treeze). In the years when the former Soviet Union was separated from the rest of the world by physical and intangible walls, many things were developed that are just now coming to light. One of these developments is TRIZ. Genrich Altshuller, a Russian patent-office employee, developed it as a system for categorizing and identifying new concepts and designs submitted for patent registration.

Altshuller reasoned that a new design can only be classified into a few dozen categories. He also believed that every design idea has conflicting issues and problems. When a solution is found to one problem, the design is often weakened by introducing other problems or limits. Altshuller named this methodology the "Theory of Inventive Problem Solving" (translated from Russian as the acronym TRIZ).

Altshuller termed the conflicts in design "contradictions" where an improvement made in one feature leads to the deterioration of another. For example, a contradiction in designing light bulbs is that if energy efficiency is improved, the life span of the bulb is shortened. Conversely, removing a contradiction means creating a better situation without accepting a trade-off. In the case of developing better light bulbs, high temperature ensures efficiency and low temperatures ensure longer bulb life. Altshuller would have stated the problem like this: "How can the temperature be both high and low?" Why should this be important? If the design concepts are tested against the TRIZ lists, it can help determine if the best alternative has been selected. Table 6-4 summarizes TRIZ.

QUALITY MAPPING

Hand-in-hand with Value Stream Mapping is quality mapping, which helps make visible what would otherwise be hidden. Quality mapping forces key questions to be asked about every process, no matter how entrenched the current version.

By developing a grid or table on a flip chart (or a roll of wrapping paper), findings can be documented to identify opportunities for improvement. The process is to ask the same quality-related

Table 6-4. TRIZ summary

Relationship to Six Sigma	Helps identify design conflicts
Who needs and uses it	Black belts, green belts, engineers
Cost	Moderate
Strengths	Tests design assumptions against a common set of criteria
Limitations	Requires critical thinking and time to process the information or test ideas
Process complexity	Medium
Implementation time	1 week–3 months
Additional resources	See Bibliography
Internet search key words	TRIZ, inventive problem solving
Internet URLs	www.mazur.net/triz/ www.aitriz.org/ www.amazon.com/exec/obidos/ASIN/0964074036

questions for each process, asking not just what defects are generated at each station, but what defects are found at each station. This process sets up an opportunity to perform a cause-and-effect diagram for each set of defects, generally resulting in dozens of ideas for improvement. Although the example in Table 6-5 is relatively small, quality maps often end up covering entire walls for more complex processes.

Quality mapping is summarized in Table 6-6.

STATISTICAL ANALYSIS TOOLS

The list of higher-order statistical tools is long. Operators and experienced team leaders most often are experienced with using some of the simpler, more mundane statistical tools (for example, Pareto analysis, Ishikawa diagram, scatter diagram, etc.) for analysis. However, at some point, there may be a need to delve deeper into the statistical toolbox. There are many community-college courses and on-line distance-learning opportunities when the need for additional training becomes apparent.

Table 6-5. Quality map

	Shear	Punch	Sand	Form
Quality specification	Print	ACME flat pattern	S-305	Customer print
Inspection frequency	First and last part	First and last part, sampling plan	100%	Sampling plan
Inspection method	Tape measure, thickness gage	Template calipers, visual	Visual/samples	Calipers/protractor
Defect potential of this process	Wrong size, wrong quantity, wrong alloy	Slug marks, burrs	Over-sanding direction, excessive burrs	Bend direction, wrong angle, die marks
Potential causes	Lack of training	Dull tooling		Operator error, burrs on dies
Common defects found during this process	Mismarked material	Tool maintenance		Slug mark, grain direction
Ideas for improvement		Third shift tool-maintenance program		Die maintenance

Table 6-6. Quality mapping summary

Relationship to Six Sigma	More of a visual tool, quality mapping is easy for new teams to use. It helps make learning a process much easier.
Who needs and uses it	Project teams
Cost	Low
Strengths	Quick to use and implement, quality mapping helps build a strong foundation for teams to apply other tools.
Limitations	Can be opinion (subjective) rather than data (objective) driven if not well managed
Process complexity	Easy
Implementation time	1 day–1 week
Additional resources	See Bibliography
Internet search key words	Continuous improvement
Internet URLs	www.ua.edu/advancement/cqi/cqitext.html

The challenge for any continuous-improvement disciple is that the time when the knowledge is needed may not necessarily be the time when he or she can afford to take time to learn. Even having access to a software package that performs the mathematical functions does little good if the results of the calculations mean nothing to the team. Someone within the organization must take the responsibility to learn the fundamentals of higher-order statistical techniques if powerful analysis tools like regression analysis, correlation analysis, and others are to be used.

Other important statistical tools that are not discussed in this text include:

- process capability studies (C_p and C_{pk}),
- hypothesis tests,
- regression analysis,
- correlation analysis,
- Chi Square,
- T-test,

- paired T-test,
- gage repeatability and reliability,
- run charts, and
- analysis of variation (ANOVA).

Higher-order statistical techniques are summarized in Table 6-7. Visit some of the recommended websites for free or low-cost statistical software demos and greater detail about application of these tools.

Table 6-7. Higher-order statistical techniques summary

Relationship to Six Sigma	Valuable tools in comparing data collected during Design of Experiments and other tests
Who needs and uses them	Black belts, green belts
Cost	Low (except for time and training)
Strengths	Helps to confirm test results
Limitations	Requires data to be collected and interpreted, and time to run tests
Process complexity	Medium to high
Implementation time	1 week–1 year, depending on experience of teams
Additional resources	See Bibliography
Internet search key words	statistics, statistical analysis tools
Internet URLs	www.analyse-it.com/default.asp www.psych.nmsu.edu/regression/home.html www.asq.org/ www.amstat.org/

CONCLUSION

After completing the Analyze step, the results should provide the steering team or sponsor a clear understanding of the potential causes for defects and a definition of why particular causes were chosen to investigate. Evidence in the form of data should be available to help anyone understand what led to the recommendations.

The material should be consolidated into a clear an understandable interpretation with little possibility for argument. In a court case, the forensic scientist must come prepared to show strong evidence a crime was committed or not committed. A similar approach should be taken to provide a strong case to improve one or more elements of the process or else show evidence that the process needs no modification.

7

Step Four: Improve

The rubber meets the road in the improvement step. All the measuring and analyzing in the world is of value only if the opportunity for improvement is identified and capitalized on by selecting and applying some of the tools presented here. Through the use of failure mode and effects analysis, Kaizen, Kaikaku, quality circles, Just-In-Time and set-up reduction, this chapter explores how teams can use the synergistic power of collective human minds to solve virtually any problem.

Discussions of 5-S, Kanban, cellularization, automation, and pull systems explain how even job shops can improve the flow of product and quality information, while reducing inventories. A section on team development will cover coaching, idea generation, helping teams manage their meetings effectively, and the importance of fostering a "learning" enterprise. Equally important is the teams' ability to brainstorm and help others adapt to change.

The five steps to effective corrective action are not complicated, but often are overlooked in favor of dealing with problems in a more cursory fashion. The root problems are, therefore, never identified and will resurface repeatedly. When used conscientiously, the steps to effective corrective action will reduce the chances for recurrence.

FAILURE MODE AND EFFECTS ANALYSIS

Failure Mode and Effects Analysis (FMEA) is a potent tool for focusing team activities on critical inputs and variables in the process under study. It helps establish priorities and can help a team avoid spending time collecting data on a non-critical aspect of the process. It also draws attention to the interrelationships between inputs and variables. Most importantly, it assigns a level of risk to each variable.

FMEA is used to:

- Identify potential failure modes (how the process could generate a defect).
- Identify the effect of a defect (severity of a failure and its potential effect).
- Identify potential causes (rate the probability of a failure).
- Calculate the ability of an operator to detect a failure.
- Multiply the three risk values (severity, likelihood, and detection difficulty) together to determine a risk priority number (RPN).
- Select the highest RPN.
- Brainstorm ways to eliminate or reduce the risk of failure.

When assigning a level of severity, occurrence, and detection to the worksheet, a common set of criteria is used such as presented in the example shown in Table 7-1.

Assigning a level of risk associated with each kind of occurrence is as important as identifying the occurrence itself. Multiplying the number assigned to the severity of an occurrence by the number assigned to its likelihood and then by the ease or difficulty in detecting the defect yields the risk priority assessment. A higher number indicates an increase in risk.

In the example shown in Table 7-2, grinding errors are quickly obvious, while the weld defect is far less obvious. Therefore it has a much higher RPN. It should be clear to the team where they need to focus their efforts.

Failure Mode and Effects Analysis is summarized in Table 7-3.

KAIZEN AND KAIKAKU

Kaizen—the term and the process—is all about making things better. Literally meaning continual, incremental improvement, this technique can be used to improve everything from work-order entry to set-up reduction, from engineering processes to packing and shipping practices. Kaizen can be further defined as a set of team-based problem-solving practices that support the major systems of Total Quality Management (TQM), Just-In-Time (JIT),

Table 7-1. Failure Mode and Effects Analysis

	Rating	Severity A failure could cause:	Occurrence Failure could happen:	Detection Potential failures are:
Bad ⇑	10	Injury	Once per day	Not detectable
	9	Liability issues	Once per week	Occasionally detectable
	8	Product uselessness	Once per week	Systematically sampled
	7	Angry customers	Once per month	100% inspected for
	6	Major malfunction	Once per quarter (3 months)	Reduced by Poka-Yoke
⇓ Good	5	Customer complaints	Twice per year	Monitored for by statistical process control (SPC)
	4	Minor performance loss	Once per year	Prevented by use of SPC
	3	Nuisance	Once every 3 years	100% checked (key features)
	2	Minor negative effects	Once every 5 years	Automated inspection
	1	Unnoticeable effects	Once every 25 years	Defect is obvious

Table 7-2. Risk priority analysis based on FMEA

Process	Potential Failure Mode	Effect of Failure	Severity	Potential Causes	Occurrence	Current controls	Detection	RPN	Corrective Action	Assignment and Target Date for Completion
Welding	Cold weld	Weld failure	9	Operator error	7	None	9	567		
Grinding	Over-grind weld	Weld failure	9	Operator error	7		1	63		

Table 7-3. Failure Mode and Effects Analysis (FMEA) summary

Relationship to Six Sigma	Great tool for identifying reasons and risks of defect occurrences
Who needs and uses it	Black belts, green belts, team leaders
Cost	Low
Strengths	Quantifies what might otherwise be an opinion and focuses teams on the low-hanging fruit
Limitations	FMEA does not fix anything. It simply identifies and ranks the opportunity.
Process complexity	Easy
Implementation time	1–4 weeks
Additional resources	See Bibliography
Internet search key words	FMEA, Failure Mode and Effects Analysis
Internet URLs	www.reseng.com/fmea/index.htm www.qualitycoach.net/fmea.htm

Total Product Maintenance (TPM), policy deployment, suggestion systems, and value-stream activities.

Much like the Define, Measure, Analyze, Improve, and Control (DMAIC) elements of the Six-Sigma system, there are steps involved here that mirror a similar approach:

1. Select a project.
2. Understand the current situations and define the objectives.
3. Analyze data to identify root causes.
4. Establish countermeasures.
5. Implement countermeasures.
6. Confirm the effect.
7. Standardize procedures.
8. Review results.

Most explanations of Kaizen reference the Plan-Do-Study-Act/Standardize-Do-Check-Act (PDSA/SDCA) cycles. These techniques are key to the thought process and the never-ending cycle of improvement. The PDSA cycle is shown conceptually in Figure 7-1.

As soon as an improvement is made, the PDSA cycle begins all over again. The SDCA cycle, as shown in Figure 7-2, is much like

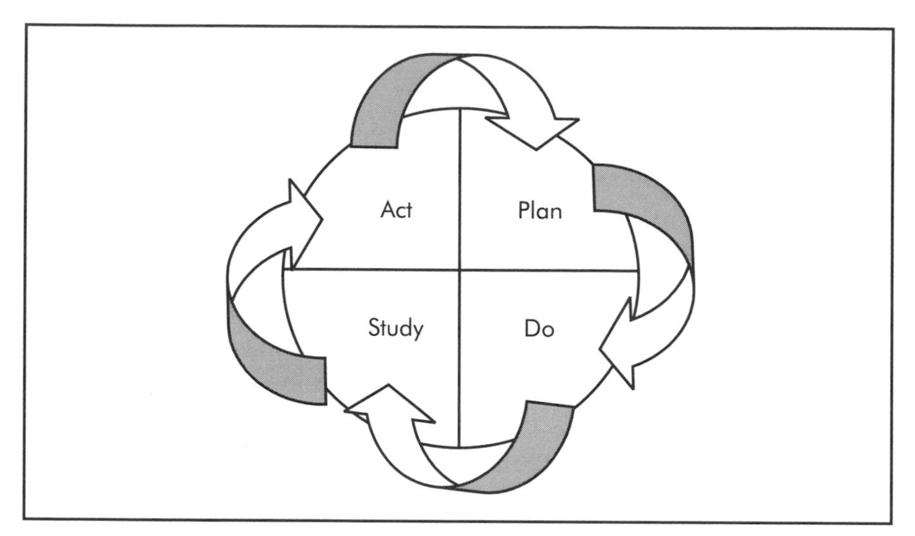

Figure 7-1. Plan-Do-Study-Act cycle.

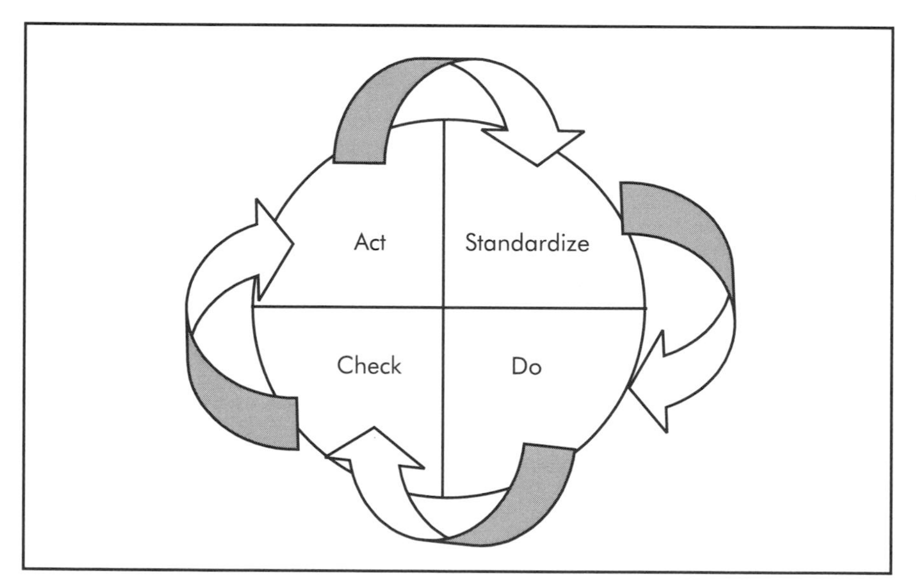

Figure 7-2. Standardize-Do-Check-Act cycle.

the PDSA cycle, but focuses on sustaining gains made during the improvement process. Here, *standardizing* means to build it into the fabric of the team's activities: changing behaviors, then *doing* (implementing the plan). *Checking* may take the form of auditing the process from time to time to verify that the improvement has been sustained. If not, corrective action is taken. The SDCA cycle is usually performed (*act*) after debugging the new process, when the team has been allowed time to accustom themselves to the new behavior.

The power of Kaizen is recognized only if everybody at every level in the organization is observant of the opportunities to remove *muda*, the Japanese word for waste. Waste is everywhere. Just as you must take out the garbage regularly or risk cluttering or even stinking up the house, you need to develop the eyes, ears, and a nose for waste at work. When you do, you will then have a never-ending list of Kaizen projects to work on.

A typical Kaizen event might last from a couple of hours to a week, depending on the scope of the project. Any project larger than what a team can tackle in a week should be separated into several more manageable projects.

The following list represents the team activities for a typical week-long Kaizen project.

- Pre-Kaizen preparation: prepare training materials; select the team; define project scope and objectives; collect data and other information important to the team; plan facilities, logistics, and lunches; and arrange the Kaizen tool kit (stopwatches, calculators, tape, flipcharts, etc.).
- Monday: train team; identify problems; define the objective statement or goal; and observe the process (collect operator cycle times, etc.).
- Tuesday–Thursday: brainstorm solutions; test ideas; simulate solutions; modify solutions as required; develop new standard work procedures (work instructions); cost-justify recommendations; determine effective ways to ensure sustainability; and meet with management for updates each afternoon.
- Friday: give team presentation; have celebration lunch; and turn over Kaizen file to value-stream leadership.
- Post-Kaizen activity: complete 30-day list assignments and audit.

During the Kaizen event, the team will be using some of the critical-thinking and problem-solving tools, and the results can be remarkable.

As discussed previously in the discussion on the Theory of Constraints, there is a tendency to jump to solutions before the question (or the real problem) is known. For example, one company wanted to perform a Kaizen event on their manufacturing process. Their operators were spending too much time performing calculations at the machine. Parts languished while machine operators studied blueprints and part drawings.

By asking a couple of key questions, it was discovered that the real constraint was not in the fabrication system, but in the engineering and drafting processes. The operators had little confidence in the accuracy of the prints, and the part drawings were delivered to the shop in a format that required the machine operators to calculate many of the machine settings themselves.

After a three-day Kaizen event held with the engineering department, this company went from delivering $80,000 per week (sales value) in engineered shop drawings to $137,000 a week. The quality problems were eliminated by implementing a buddy-check system, which will be discussed later.

Kaizen is the best vehicle for improving processes like machine set-up, quality mapping (and waste reduction), workplace organization, and defining standard work—all of which will be discussed shortly.

Often referred to as "five days and one night," Kaizen weeks can be long and exhausting. The fact that people cannot generally sleep well during this week is related to the dynamic energy created by being a part of something new, and the human mind's ability to lock into a creative mode and its inability to turn it off at night. Participants should be told about this early, possibly the week before the event so they can schedule their week to allow for what might be long days and short nights. One helpful technique is to place a notebook and pencil on the nightstand during Kaizen weeks. This allows you to write down any idea that occurs during the night. There are many rules necessary for conducting an effective event. The basic *rules of Kaizen* are:

- Understand the process.
- Keep an open mind.
- Maintain a positive attitude.
- Never leave in silent disagreement.

- Create a safe environment.
- Practice mutual respect.
- Treat others as you would like to be treated.
- Never exploit position or rank (one person, one voice).
- There is no such thing as a dumb question.
- Just do it!

Kaikaku

Kaikaku is the big brother of Kaizen. Closely associated with the English term for radical change, this event might be used where an entire manufacturing plant is being redesigned to allow better flow or when an entire engineering department is looking to change the way they receive, develop, exchange, and store information. It may require a number of Kaizen events to follow-up on the accomplishments and the level of changes demanded by a Kaikaku scope statement.

The tools of Kaizen and Kaikaku are much the same; Kaikaku just happens on a much larger scale. One great example of Kaikaku was implemented at a secondary wood-component manufacturer, where the goal of switching from departmentalized operation to value-stream operations meant complete restructuring of the leadership team. For close to 30 years, they had been horizontally aligned by department, as shown in Figure 7-3.

Now, they were reassigned to value streams where the responsibility reached across all departments (Figure 7-4). Such a radical change certainly qualified as a Kaikaku event.

The change not only affected those in the production-supervision area, but also incited many changes by the person responsible for managing the computer systems used for planning. Re-programming and physical relocation of some equipment took place within a few weeks of the event. For the most part, the value streams continued to operate at a distance. Thanks to Kaikaku, the emotional connection was made among all parties, however, they were still working toward physically co-locating elements of the value streams for some time afterwards.

Kaikaku and Kaizen are powerful tools for implementing change, but they require commitments to follow-up. Otherwise, the efforts are easily diluted by the everyday activities that seem

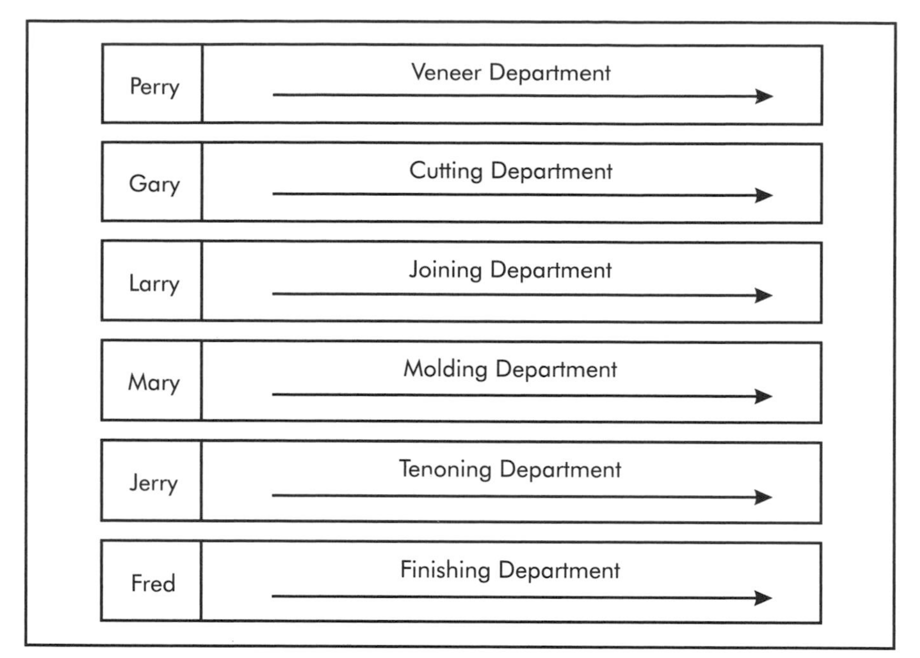

Figure 7-3. Horizontally aligned departmentalization.

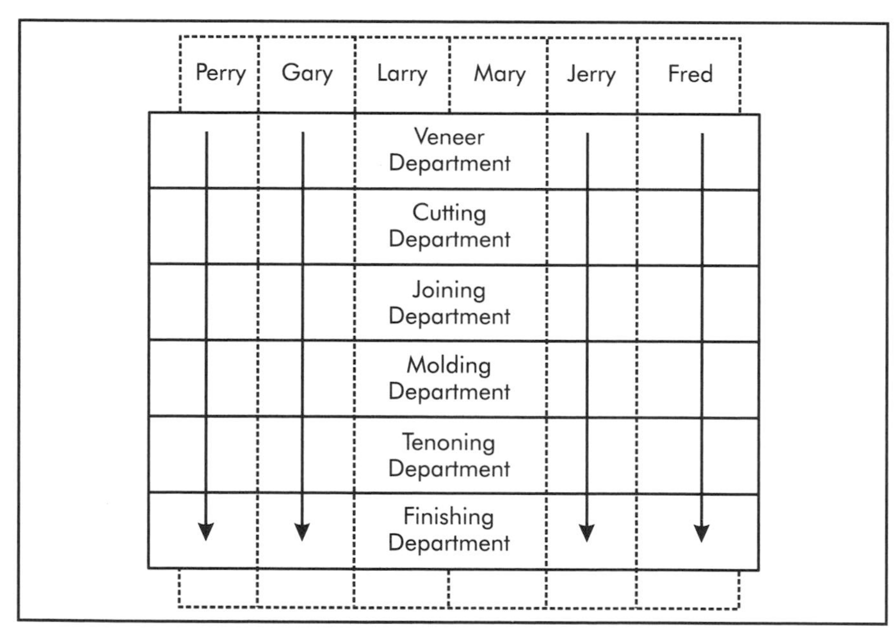

Figure 7-4. Vertically aligned departments after Kaikaku event.

to encroach into the forefront. Like the real-estate mantra, "location, location, location," for Kaizen, the rule is "execution, execution, execution." You must follow through if you intend to reap the rewards of Kaizen and Kaikaku. Kaizen and Kaikaku are summarized in Table 7-4.

Table 7-4. Kaizen and Kaikaku summary

Relationship to Six Sigma	Fundamental tools of change
Who needs and uses them	Black belts, green belts, value-stream leaders, process owners, teams
Cost	Moderate—event-based Kaizens usually cost about $5,000 per week for in-house events
Strengths	Powerful tools for implementing change quickly
Limitations	Very few limitations
Process complexity	Easy to moderate, depending on the scope of the project
Implementation time	1–30 days
Additional resources	See Bibliography
Internet search key words	Kaizen, Kaikaku
Internet URLs	www.productivityinc.com www.kaizen-institute.com/ gchapman.cbpa.ewu.edu/Acctg357/ webpages/Kaizen/Kaizen.html

JUST-IN-TIME

Just-In-Time (JIT) is one of the pillars in the Toyota Production System. Lead by Taiichi Ohno and the team at Toyota, the JIT transformation of auto manufacturers worldwide has been nothing less than life changing for many companies, teams, and individuals. This transformation has not only affected assembly lines at car makers, but has moved upstream and downstream to parts suppliers, showrooms, service garages, and repair facilities as well as raw-material makers and suppliers. The face of the manufacturing world has been changed forever due to the creativity and hard work of Ohno and his team.

The focus is on the elimination of waste in all its forms. No company will ever eliminate all waste, so this journey never ends. However, the focus must be maintained or the journey trails off as people lose track of the objective. Coupling JIT with Six Sigma will help avoid evaporation of gains.

Often, the terms JIT and Kanban are used interchangeably. However, Kanban is a tool within the JIT toolbox. JIT really is about timing production flow to the actual demand of the customer. This is easier to see and manage within an original equipment manufacturer (OEM) environment. Nonetheless, the tools can apply in engineer-to-order and make-to-order settings as well.

Just-In-Time deliveries to a customer might be more difficult for the job shop to manage, because what is going to be ordered by the customer is not really known (except where a blanket order exists). Yet, once something is ordered, can a job shop deliver it Just-In-Time to the scheduled delivery?

In an example of a sheet-metal shop, can the shear deliver blanks to the punching department Just-In-Time? Can the punch provide the processed blanks to the press brake Just-In-Time for it to bend the parts?

The goal of JIT is to reduce throughput time and, in the process, work-in-process (WIP) inventory is reduced. There are very few distinctions between JIT and what has been titled the Lean approach. JIT tends to focus on the shop, while Six Sigma and Lean take a more holistic approach to improving the entire enterprise.

Here are the "Ten Commandments of JIT" (Hirano 1989):

- Throw out old, tired production methods and concepts.
- Don't think of reasons why it won't work; think of ways to make it work.
- Don't make excuses, just deal with current conditions.
- Don't wait for perfection; 50% is fine for starters.
- Correct mistakes immediately.
- Improvements should not be costly.
- Wisdom arises from difficulties.
- Ask why at least five times until you find the real cause.

- Better the "wisdom" of ten people than the "knowledge" of one.
- Improvements are unlimited.

The tools associated with the JIT system are the same ones described throughout this text. The fact that these tools dovetail together so well with other systems supports the fact that this is not an either/or choice. The many tools available complement rather than conflict with each other. The JIT toolbox includes:

- 5-S (discussed later);
- pull systems;
- multi-process operations, multi-tasking operators;
- Kanban (signal to move or make);
- visual control;
- leveling (level loading);
- standard operations (standard work);
- Jidoka/Poka-Yoke;
- safety and maintenance;
- total quality control;
- supplier partnerships;
- process flow;
- set-up reduction;
- synchronous production; and
- employee involvement.

The methodologies of JIT mirror many of the process steps outlined in the discussions of Six Sigma, Lean, and other methodologies:

1. Select a pilot project.
2. Form a team.
3. Study the existing process.
4. Prioritize the opportunities.
5. Generate ideas for improvement.
6. Document the new process.
7. Train people in the new process.
8. Then, repeat the process.

Elements of the DMAIC process can be discerned within some of these steps. The Shewhart cycle of Plan-Do-Study-Act (PDSA) is also recognizable. Just-In-Time is summarized in Table 7-5.

Table 7-5. Just-In-Time summary

Relationship to Six Sigma	The fundamental tools of JIT apply to any improvement program, including Six Sigma.
Who needs and uses it	Black belts, green belts, value-stream managers, team leaders
Cost	Moderate
Strengths	Wide array of tools within one package
Limitations	Takes time to apply the many tools
Process complexity	Difficult
Implementation time	6 months–3 years
Additional resources	See Bibliography
Internet search key words	JIT, Just-In-Time
Internet URLs	www.ashland.edu/~rjacobs/m503jit.html www.industryweek.com www.ece.curtin.edu.au/~clive/jit/jit.htm

SINGLE-MINUTE EXCHANGE OF DIES AND SET-UP REDUCTION

Single-Minute Exchange of Dies (SMED) is the goal of changeover from one kind of product to another. One minute or less is the target. Single-Digit Exchange of Dies (SDED) is a slightly less aggressive goal. It is often used as a stepping stone to greater gains by job shops and engineer-to-order shops where the goal is less than 10-minute set-up times. Even the single-digit goal raises some people's eyebrows when teams first begin working. They often doubt the possibility of taking 40, 50, or 60-minute set-ups down to 10 minutes or less in just a Kaizen event or two. A team of motivated, open-minded, creative people rarely fail to cut set-up times in half on the first pass, and then by 30–50% again on a second effort.

Set-up reduction has a close relationship with the application of standard work and best practices. From a quality perspective, this will have a secondary benefit of helping attain Six-Sigma levels of performance. This is because in most shops, current practice allows each operator to design or modify the set-up to suit his or her preferences. In one case, operators on a swing shift came in and completely adjusted a machine that had been set up and running fine all day, simply to suit their own preferences. Not only can this negatively affect productivity, but also quality and consistency go out the window.

Creating set-up sheets or best practices for the thousands of part numbers processed by a job shop might seem an expensive proposition. Yet, by focusing on the 80/20 rule, generally all the primary part numbers can be documented in a month or two. Developing a form where the operators can simply check a box as compared to writing redundant information over and over again will speed this process. Figure 7-5 shows an example of a set-up form.

The steps of SMED are:

1. Observe (videotape) the set-up.
2. Document operator set-up time (by activity).
3. Measure set-up distance (using a spaghetti diagram).
4. Define internal/external elements.
5. Separate external elements from the set-up.
6. Shift as many internal elements to external as possible.
7. Streamline internal elements.
8. Streamline external elements.

For example, the average press brake can take upwards of an hour to set-up. *Set-up* is the time from the last good part to the first good part of another run. Looking at the procedures involved, a list will look something like this:

- Log off last job, log into new job.
- Pull out tooling.
- Study blueprint.
- Find the new tooling.
- Put the last job away.
- Get the new material for the next job.
- Study the blueprint.

Hardware Set-up Sheet		
Operator Name:		Date:
Customer (& Division:)		
Part Name:		
Customer part #:		Rev: / / /
ACME Part #:		Rev: / / /
Set-up #1	Set-up #2 (if applicable)	
Machine #:	Machine #:	
Hardware type:	Hardware type :	
❑ Auto feed	❑ Auto feed	
❑ 'J' Frame (toolholder)	❑ 'J' Frame (toolholder)	
❑ Special tool	❑ Special tool	
❑ Anvil # (toolholder)	❑ Anvil # (toolholder)	
Hydraulic pressure: psi ❑ Run ❑ Set-up	Hydraulic Pressure: psi ❑ Run ❑ Set-up	
Top tool #:	Top tool #:	
Bottom tool #:	Bottom tool #:	
Sketch of set-up is required		
Doc # F9.02 Rev 05/16/02	Page 1 of 1	Approval:

Figure 7-5. Example of a machine set-up form.

- Set tools in the right place.
- Tighten tooling.
- Get special back gages if needed.
- Move back-gage stops.
- Program numerical control (NC).
- Test bend the part.
- Adjust program.
- Run.

Examining the time taken for each of these steps, the total duration is 35 minutes. Split these steps into external (involving people other than the operator) and internal (operator only). Then, ask a series of questions for each of the external activities.

- "Is Dan (the operator) the only person who can log himself on and off jobs, or is there a member of the pit crew who could log him in and out so that he could focus on the set-up?" Hmm, yes, maybe someone else could log him in and out.
- "How about pulling the tooling, can anyone else do that?" Well, no, otherwise what would Dan do?
- "What if there was a set-up sheet that told Dan what tools to get? Would he still need to spend time studying that print?" Hey, maybe someone else could study the blueprint or set-up sheet and pull the new tools for Dan. Someone could also put the last job away and get the new material. He or she could also get the special back gages ahead of time.

So this process goes, and soon some major reduction possibilities have been discovered even before brainstorming ideas for streamlining the internal process steps.

If six set-ups are performed in a day and 85.8 minutes of machine downtime can be saved every day, not only does that relieve what may be a bottleneck, but it translates to nearly 360 hours per year or eight weeks. Eight weeks of capacity has just been gained in addition to doing more set-ups for less cost. Now lot sizes can be reduced.

Next, looking at the internal steps, the operator is asked for ideas. "What would it take to cut the time for this activity in half?" Then, ideas like, "If I had an air wrench" or "If we moved that rack over here" or "If we used NC tapes instead of programming each part by hand" begin to surface. At the end of this session, it's a matter of testing the ideas, simulating the new approach, and documenting the new set-up process so that it is useful not just for one part, but for all parts.

Other SMED techniques include:

- Preset desired settings.
- Use quick fasteners (quarter-turn screws, special clamps).
- Use locator pins (tapered pins, shot pins).
- Use Poka-Yoke devices to prevent misalignments.
- Eliminate hand tools.
- Make movements easier.

Applying the principles of Six Sigma to the activity of set-up reduction can provide long-lasting gains. Here's how Six Sigma can be used to establish measurable and improvable objectives.

1. Define: A team examines lead times and finds a certain machine with excessive set-up times (40 minutes), which is forcing the company to run larger lot sizes (for example, one-month lot sizes). This, in turn, results in producing unneeded parts and storing them until they are needed. The goal of the team is to reduce set-up time by 75% (to 10 minutes). This will allow lot sizes to be reduced to weekly batches instead of monthly batches (at no additional cost).
2. Measure: The team measures and records each element of the set-up as described in the SMED process. The goal is established, deemed realistic, and a means to record and make the results visible is developed.
3. Analyze: The team analyzes the set-up to establish those activities that are internal and those that are external. Each activity is identified as either value added or non-value added. A project board is provided near the machine to show the result of this analysis. A flip chart is also available where people can record their ideas.
4. Improve: The team problem-solves and brainstorms ideas and solutions to streamline both the internal and external activities. Items like multiple adjustments and multiple turns of a wrench are reduced or eliminated by employing techniques like standard die height and quick-release clamps. The team

simulates or tests the ideas for improvement, including observing and timing the improved process.

5. Control: The team develops a standard work sheet that clearly defines the best practices. New procedures are developed. A control chart that allows the operator to record the actual set-up time along with the target provides a visual representation of how closely the team is able to maintain the improvements over time.

Just like measuring the features on a physical part, the 10-minute goal should be something that is measured and recorded on a run chart, and not just once per week, or whenever the continuous-improvement team comes by to audit the process.

Obviously, Six Sigma is about controlling variation, so a "miss" toward the improved side of the scale is seen just as negatively as a "miss" toward the negative side of the set-up scale. To reduce the possibility of seeing an improvement as a negative result, each measurement that meets the target or improves the target positively could be entered as zero, and everything greater than the goal entered as the difference between the target and the actual.

For example, 25 set-ups were made with the following times: 10, 9, 12, 4, 7, 10, 10, 10, 8, 7, 4, 13, 10, 14, 6, 6, 7, 4, 6, 7, 5, 6, 9, 10, 13. The average for this group of data is: 8.28. If the target were 10 minutes, and everything under the target is valued as a zero, then these values should be entered into the Six-Sigma equation as: 0, 0, 2, 0, 0, 0, 0, 0, 0, 0, 0, 3, 0, 4, 0, 0, 0, 0, 0, 0, 0, 0, 0, 0, 3. The results are 1.159 per sigma (3.477 for 3 sigma). This means that a variation of less than 3.5 minutes can be expected for any set-up performed. If the team objective is to have the average set-up be less than 10 minutes and to vary (range) by no more than 5 minutes, then there is a fairly good chance of meeting that objective according to the results.

Finding a way to measure and control every critical activity (like set-up reduction) is the new job of the supervisor (manager). Where in the past everyone was told what to do, when to do it, and how long to take getting it done, now the data tells the operator or process owner what to expect and what is expected. SMED is summarized in Table 7-6.

Table 7-6. Single-Minute Exchange of Dies (SMED) summary

Relationship to Six Sigma	Excellent way to teach DMAIC methodology because it works on any process
Who needs and uses it	Everybody
Cost	Low
Strengths	Allows quick gains and attacks bottlenecks and constraints
Limitations	None
Process complexity	Easy
Implementation time	1 day–1 week
Additional resources	See Bibliography
Internet search key words	SMED, quick-change tooling, fast set-up techniques
Internet URLs	www.iit.edu/~ipro024/ membres.lycos.fr/hconline/engineer_us.htm

5-S

Japanese put into words what to many of us seems like just common sense. As the saying goes, "the trouble with common sense is that it is not all that common." Many struggle with what some people can do intuitively, and this is certainly the case with workplace organization.

The 5-Ss were originally developed in postwar Japan in an effort to improve working conditions and efficiencies. They are:

1. *Seire*: to sort—get rid of what is not needed.
2. *Seiton*: to set in order—organize what is needed.
3. *Seiso*: to shine—clean and inspect for potential problems.
4. *Seiketsu*: to standardize—create standards and make them visual.
5. *Shitsuke*: to sustain—ensure improvement is sustained through training.

Other interpretations of these Japanese words include:

- sort,
- systematize,
- sweep,

- standardize, and
- self-discipline.

Some argue that 5-S is the best place to start a transformation, because if you rid what is not needed and clean up what's left, you already have begun the process of waste elimination. By doing this first, rework is avoided later. Balance is key. Using all these tools in harmony is the best way to avoid false steps and false starts.

A thorough 5-S team training program should include instruction on how to:

- Prepare for an event.
- Prepare for the red-tag process (identification).
- Determine what is really needed ("When in doubt, move it out!").
- Set up temporary holding areas.
- Remove or relocate nonessential materials and equipment.
- Perform an effective workplace evaluation.
- Describe the current condition.
- Brainstorm ideas for improvement.
- Make it obvious where things belong (outlines on the floor, shadow boards).
- Determine cleaning methods and schedules.
- Replace broken or worn items, hoses, belts, tubing, etc.
- Determine proper levels of inventory.
- Identify tools that belong in the work area.
- Establish policies for storing hazardous or fragile materials and tools.
- Ensure safety policies are easy to follow and that safety is enhanced.
- Develop systems to make the work steps visual.
- Build weekly 5-S activities into the job description (standard work).
- Develop a communication board and perform audits that confirm improvements.
- Develop training materials for all employees (including new team members).

5-S can be summarized by the phrase, "a place for everything, and everything in its place."

5-S is covered in Table 7-7.

Table 7-7. 5-S summary

Relationship to Six Sigma	Quality is in the eye of the beholder. When a shop looks clean and has the perception of quality, motivation to produce higher quality is enhanced.
Who needs and uses it	Everyone
Cost	Low
Strengths	Helps bring a sense of order and organization to any activity and supports SMED.
Limitations	Does not by itself decrease the opportunity for defects
Process complexity	Medium
Implementation time	1–4 weeks
Additional resources	See Bibliography
Internet search key words	5-S, five S
Internet URLs	www.productivityinc.com/index.html www.php.co.jp/japaninface/videos/2IND5s.html www.graphicproducts.com/tech/five_s/index.htm

KANBAN

Kanban is possibly one of the most misunderstood and misinterpreted tools in the toolbox. Customers who consider themselves to be on the path to becoming Lean often arm-twist their suppliers into maintaining inordinate amounts of finished-goods inventory and refer to it as a Kanban. It definitely helps the bottom line of the customer who gets to move the responsibility and cost of storage to someone else, but is this really a Kanban?

Kanban and in-process inventory do have a close relationship, but is Kanban the goal or a tool? Are there instances where Kanban does not make economic sense? How well do Kanban systems work in pure make-to-order environments?

Let's start with the last question first. Kanban is really a signal to make something or move something. In a job shop, the only reason something is made is that there is an order for it or a good reason to expect an order shortly. Trying to maintain WIP inventory for 3,500 part numbers in the hopes of selling one is futile.

So how does Kanban apply to the job shop? This question is answered by first looking at Kanban in the traditional sense. Then, the job-shop version of Kanban is explained.

Traditional Kanban

The use of Kanban is not new. Its development grew from inventory-management techniques in which there were statistically sound reorder points. When the level of a product sitting in inventory reaches a pre-determined minimum level, another order for it is released. The benefits of a reorder point or a Kanban system are: people are freed from doing mundane calculations all day long; helps reduce inventory-management costs; and materials can be managed without paying close attention to minor fluctuations in demand patterns.

Kanban and reorder-point systems do not work well when: there are volatile and unstable order patterns; the parts are difficult to store or make; or the materials are expensive.

There are a variety of Kanban types:

- A *signal kanban* is used primarily when a process requires set-up or changeover.
- A *production kanban* is used where little or no changeover is required.
- A *transport kanban* signals multiple parts to be moved at once (that is, to a production line).
- An *in-factory kanban* is used between processes.
- A *supplier kanban* generally signals orders to a vendor.

Where there is a routine order pattern and the customer acts like a true JIT customer rather than a short-term thinker (someone trying to make their numbers look better by crippling their supplier), Kanban does have its place. When there is regular consumption that can be turned into a Takt time (daily production orders), then the filling and emptying of Kanbans will work just fine.

There are several variables that need to be identified to calculate a Kanban. The manufacturing lead time must be known (that is, how long it takes parts to go from order release to being ready to use at the point where a Kanban is established). Generally, a safety-stock level is established that will insulate against problems during the transition from a traditional batch process.

The goal of any true Kanban system is to reduce safety stocks as soon as possible after gaining confidence that the program is working. When implementing a Kanban system for the first time, there is usually a lot of inventory sitting around that must be eliminated. Two kinds of Kanban cards are developed: a true Kanban card (for example, red) that is recycled as an order to make something, and a disposable Kanban card (for example, white) that is thrown away at the end of the process, thus eliminating it as an order to make the part again.

The monthly, weekly, daily, or even hourly demand must be calculated to determine Kanban size. If the Kanban is to be replenished weekly, then this is the number used in the calculation that follows. If the goal is daily or hourly replenishment, then that respective time element is used in the calculation.

Finally, if the parts are small and stored as multiples on a pallet or in a box or tote bin, the number of pieces that fit in the container needs to be established to calculate the number of Kanban cards. This assumes that the container size will always be the same.

$$K = [D \times (L + S)] \div C \qquad (7\text{-}1)$$

where:

K = Kanban quantity
D = daily output (monthly demand ÷ days worked each month)
L = lead time (throughput time + queue time + retrieve time)
S = safety stock (as little as possible for insurance) represented in time rather than units
C = box, bin, rack, or pallet capacity (holding device)

For example:

D = 60 units (in 8 hr)
L = 4 hr (0.5 day)
S = 2 days
C = 10 pieces (box)

Then:

$$[60 \times (0.5 + 2)] \div 10 = 15$$

In this example, a Kanban is set up that would have 15 boxes, with 10 units in each box.

There are hybrid examples of Kanban in nearly every industry, whether service or manufacturing. Kanban has lost some of the initial luster it had in the 1970s and 1980s when it was discovered that Kanban systems still require inventory.

Another realization that has reduced the use of Kanban is that pulling material through a shop using the domino effect of one process or department pulling from a storage location could be virtually eliminated if the processes were close-coupled and line-balanced. The velocity of material being metered into a manufacturing process (hybrid push system) in some cases was greater than if it was being pulled through dozens of Kanban locations. This decreased the non-value-added activities of maintaining storage locations and Kanban cards.

Kanban has its place, but it is not the "magic bullet" often claimed by some customers. From the customer's perspective, finished-goods inventory makes sense and looks a lot like a Kanban, but this perspective is driven by a lack of trust in the supplier's ability to produce at high velocity.

Job-shop Kanban

In an environment where there is little or no predictability about what will sell, the traditional finished goods inventory and work-in-progress Kanban levels do not work very well. An alternative is to develop a system of generic Kanban cards. They have no relationship to a specific part number. They simply control (and limit) the amount of work in front of and after each operation. If the amount of work exceeds this level, then the Kanban cards limit overproduction.

For example, assume a company has huge amounts of inventory running through a number of departments. Figure 7-6 shows the current condition.

The push arrows show how work orders and material are pushed from one department to another. The triangle symbol represents

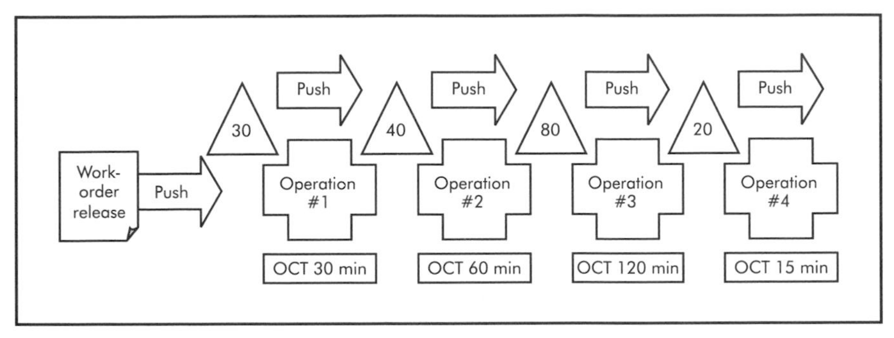

Figure 7-6. Example of a push system.

the inventory amount ahead of or behind the operation. Operation #3 is obviously the bottleneck. The operator cycle time (OCT) number under the operation signifies the time it takes to process an average lot. In this case, there are 170 lots of material in the process. Production planning is releasing jobs whenever one is sold, so the pipeline may contain more or less than 170 lots at any moment in time.

Because operation #3 is the constraint, it determines the effectiveness of the entire line. The time it takes for job #170 to make its way through the shop is simply a matter of multiplying the number of jobs by the OCT for the average job. So,

> $(30 \times 30) + (40 \times 60) + (80 \times 120) + (20 \times 15) = 13{,}200$ minutes of lead time or 220 hours, or 27.5 days (if this company is working eight-hour days)

The job-shop approach to Kanban takes out all inventory except what the operations need to have in front of them for one cycle. If Operation #3 is the pace setter, then the cycle is 120 minutes. Operation #1 needs to produce only one quarter of a lot every 120 minutes (otherwise the pipeline would fill up again). Operation #2 can produce no more than one half a lot every two hours. Operation #4 should never have more than one lot size in process.

The throughput time for any job just entering the system is now:

> $(1 \times 30) + (1 \times 60) + (1 \times 120) + (1 \times 15) = 225$ minutes or 3.75 hours

The jobs are still in the backlog and metered out by the production planner. But now the pace of the line becomes more like a heartbeat. People are focused on optimizing the entire process rather than individual operations (Figure 7-7).

If the inventory goes up for Operation #3, everyone has to stop and help rather than keep producing something that cannot be processed. To control this flow, Operation #4 becomes the Kanban signal point. Whenever a job is finished, a signal is sent back to the person responsible for releasing work orders. The pace of signaling should end up being about 120 minutes (until operation #3 is sped up). Signaling can be managed by a physical card, phone call, fax, buzzer, or other device.

By reducing inventory to the bare minimum, a lot of problems are uncovered. Be ready to deal with these immediately or else put back a little inventory to cover up the problem. If two hours are needed to solve any problem that might arise, then double the inventory safety stock.

Table 7-8 is a representation of a Kanban system set up by a team dealing with imbalances in operator cycle times. The Takt time (based on true demand) requires one assembly to be produced every 240 minutes. Because some operations process multiple units, but take far longer than one Takt time to process, the number of units in queue is greater than one. In this case, for Part A, there should never be more than 13 units in process. Part B will have as many as nine units in process, and Part C will have four or less units in some stage of manufacture.

Every 240 minutes (on average) one part comes out the end of the process into shipping, and another is introduced into the process. The total operator-cycle-time labor must be applied to each Part A, so the 13th part number just now introduced will take 13 Takt times or 52 hours (working hours only) to reach shipping. This still seems like a lot, but before the team applied Kanban, the lead time was 4½ weeks (468 hours), so this represents an 88% reduction in lead time. For this team, the financial impact of their work was staggering. Each unit was valued at about $5,000, so for every week that they reduced lead time, they saved $40,000. During the initial project, the inventory level was reduced from 14 weeks to less than eight, and this represented a $280,000 inventory reduction. In this example, the parts are pulled through.

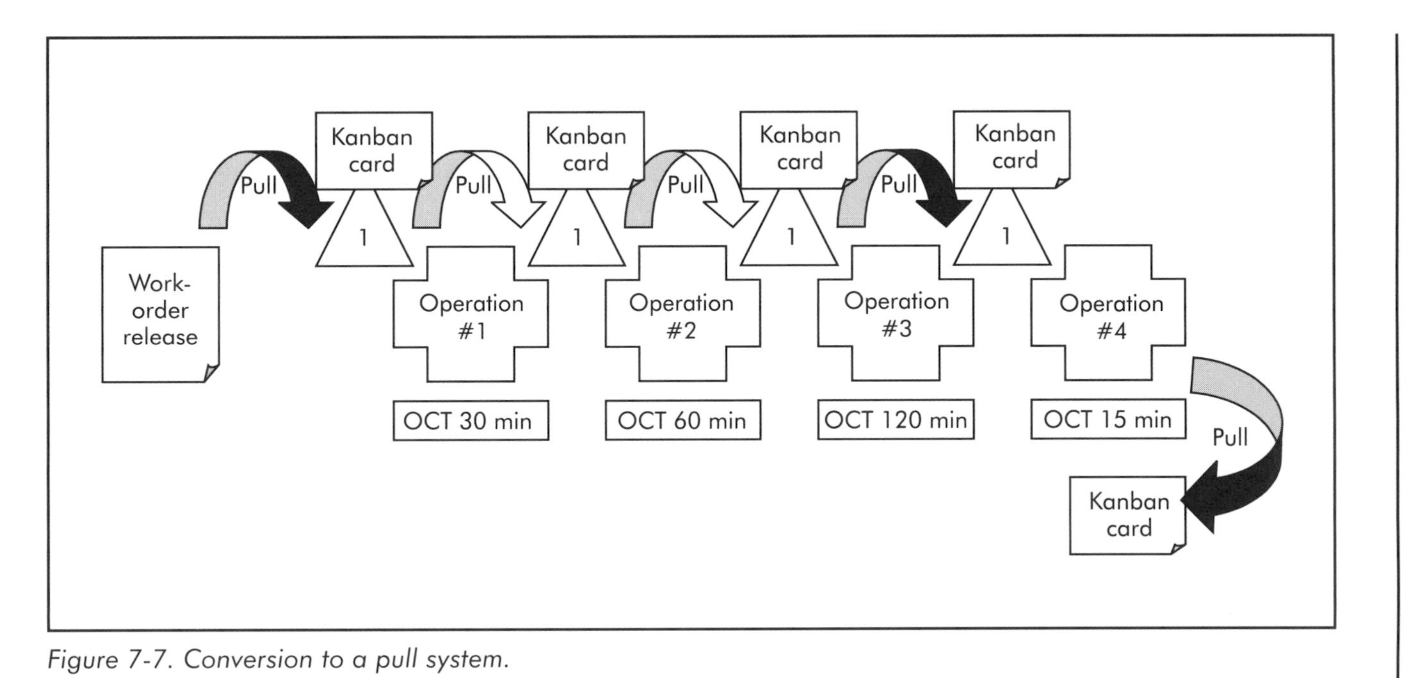

Figure 7-7. Conversion to a pull system.

Table 7-8. Takt time work sheet
for mixed model line with long operation cycles

<table>
<tr><th>Operation</th><th colspan="8">Part A</th><th colspan="4">Part B</th><th colspan="2">Part C</th></tr>
<tr><td>Step 1</td><td>13</td><td colspan="2">12</td><td colspan="2">11</td><td colspan="2">10</td><td>9</td><td colspan="2">9</td><td colspan="2">8</td><td colspan="2" rowspan="2">4</td></tr>
<tr><td>Step 2</td><td colspan="8">8</td><td colspan="4">7</td></tr>
<tr><td>Step 3</td><td colspan="8" rowspan="2">7</td><td colspan="4" rowspan="2"></td><td colspan="2" rowspan="2"></td></tr>
<tr><td>Step 4</td></tr>
<tr><td>Step 5</td><td colspan="2">6</td><td colspan="2">5</td><td colspan="2">4</td><td colspan="2">3</td><td>6</td><td>5</td><td>4</td><td>3</td><td>3</td><td>2</td></tr>
<tr><td>Step 6</td><td colspan="8" rowspan="6">2</td><td colspan="4" rowspan="6">2</td><td colspan="2" rowspan="16">1</td></tr>
<tr><td>Step 7</td></tr>
<tr><td>Step 8</td></tr>
<tr><td>Step 9</td></tr>
<tr><td>Step 10</td></tr>
<tr><td>Step 11</td></tr>
<tr><td>Step 12</td><td colspan="8" rowspan="10">1</td><td colspan="4" rowspan="10">1</td></tr>
<tr><td>Step 13</td></tr>
<tr><td>Step 14</td></tr>
<tr><td>Step 15</td></tr>
<tr><td>Step 16</td></tr>
<tr><td>Step 17</td></tr>
<tr><td>Step 18</td></tr>
<tr><td>Step 19</td></tr>
<tr><td>Step 20</td></tr>
<tr><td>Step 21</td></tr>
<tr><td></td><td colspan="9">Sunday PM</td><td colspan="5">Monday AM</td></tr>
<tr><td colspan="4"></td><td>10:30 PM</td><td>10:56 PM</td><td>11:23 PM</td><td>11:49 PM</td><td>12:16 AM</td><td>12:42 AM</td><td>1:09 AM</td><td>1:35 AM</td><td>2:02 AM</td><td>2:28 AM</td><td>2:55 AM</td></tr>
</table>

If an operator working on part #1 has no place to store part #2, then the team members stop working on part #3 until the reason for the bottleneck is flushed out.

Kanban is summarized in Table 7-9.

Table 7-9. Kanban summary

Relationship to Six Sigma	Kanban supports the principles of working on the right things at the right time. It avoids overproduction, hidden forms of waste, and part degradation due to storage.
Who needs and uses it	Production planning, value-stream managers, black belts
Cost	Moderate
Strengths	Frees up capital currently held in the form of inventory and can make problems apparent
Limitations	By making problems apparent, can cause organizational stress if done all at once
Process complexity	Medium
Implementation time	1–6 months
Additional resources	See Bibliography
Internet search key words	Kanban, inventory reduction, manufacturing pull signals
Internet URLs	www.mep.org/leanmanufacturing/html/pull_kanban.html www.atnlean.com/services/kanban.asp

PULL SYSTEMS

Flow is an essential focus in manufacturing, as has been mentioned in this text repeatedly. If you have ever been in a flood, you know how quickly an excess of water can dam up at a river's bottlenecks and create havoc. So too in manufacturing. Material should flow steadily, without the waste incurred when it flows too fast or too slow. By ensuring that no operation or department is overproducing, surges of work are avoided that alternately starve and then flood downstream operations.

In pure assembly operations, it is relatively easy to attain flow. In make-to-order shops, it is significantly more challenging, simply because the parts do not all process at the same rate through every operation. In fact, they often skip certain processes or operations.

Are there alternatives to one-piece flow (making one part at a time and passing it along) as in automotive assembly lines? Yes, one-pallet flow, one-box flow, or one truckload flow are possible. The goal should be to see the material moving rather than sitting, and in some cases one work-order flow is better than no flow at all.

Some things that will help control flow:

- Avoid process-functional layouts (departmentalization).
- Reduce line imbalances (synchronize the work to avoid inventory pile-ups).
- Reduce long set-up times on equipment (any set-up over 10 minutes is too long).
- Reduce machine breakdowns or unreliability by performing preventive maintenance.
- Measure and reward operators for preferred behavior (eliminate chronic absenteeism).
- Cross train so people can help flush out bottlenecks.
- Focus on "getting it on the truck" rather than "keeping a machine running."
- Reduce the number of defects and the need to inspect.

Japanese manufacturing firms have a saying, "The next operation is your customer." Although this is intended to relate primarily to quality, looked at from a flow perspective, it suggests how product should be delivered to your customer. Should a sale be forced upon your customer? Should he or she be forced to take your excess production? What if a restaurant did this to you? You order lunch, but they bring you dinner and breakfast as well. You can't eat it all, so you have to take it with you and store it. This is no different than forcing your internal customer to find a rack to store material that he or she cannot currently process. It may be optimizing one operation to run more than a subsequent customer can process, but it is sub-optimizing the entire operation to overproduce.

The idea of a pull system is to meter the work through the shop at the rate each downstream operation is capable of consuming it. Workers often fail to recognize that when they overproduce, they have to do extra work—like paperwork—necessary only because the material is going into a waiting area.

Stop-and-go production is like being stuck on a freeway during rush hour. If everyone on the freeway would just travel at a safe speed, traffic would flow more smoothly and stress-free, and accidents would be reduced, along with paperwork and insurance costs. Eliminating the "fender benders" in manufacturing will do a lot for overhead, stress, paperwork, and insurance costs as well.

Pull systems are summarized in Table 7-10.

Table 7-10. Pull systems

Relationship to Six Sigma	Pull systems support the Six-Sigma principles of giving customers what they want when they want it. It makes things easier to visualize, measure, and control.
Who needs and uses it	Value-stream leaders, black belts
Cost	Low
Strengths	High payback when inventory is reduced along with lead times
Limitations	Does not focus on quality issues
Process complexity	Medium
Implementation time	2–6 months
Additional resources	See Bibliography
Internet search key words	Pull systems, flow
Internet URLs	www.searchmanufacturing.com/Manufacturing/Lean/books.htm web.starlinx.com/rbvollum/whpaper.htm

CELLULAR CONFIGURATIONS

When you look at a beehive and all those individual cells of the honeycomb, it is amazing how strong this network of cells is once they are grouped together—much stronger than if you just dumped a bunch of wax in a pile. The objective of creating manufacturing cells and cell teams within an organization is to create a network of cells that have much greater flexibility, strength, and resiliency than they would functioning as individuals.

Just putting people in the same area does not make them a team. There is a need for teams to understand their individual responsibilities and learn how to manage the group dynamics and interrelationships required to operate as a manufacturing cell.

Engineers are usually analytical people and that means they like organization. Having equipment in nice straight lines and at 90-degree angles looks very orderly and makes most engineers comfortable and happy. Yet, machines neatly aligned at perfect right angles to each other do not necessarily make for a productive working environment.

In one case, a work cell was set up where the press brake operator was also responsible for setting up and monitoring a robotic press brake. The maintenance team tried to dictate to the cell team that they should leave 6 ft (1.8 m) of space between press brake machines in case of a breakdown. When asked how often a breakdown actually occurred, the answer was, "Oh, every year or so." To facilitate a once-a-year event, the maintenance team wanted to force the operator to walk an extra 6 ft (1.8 m) from machine to machine every day, year round. Figuring four trips per hour, this added up to 96,000 ft (about 18 miles) (29,261 m or 29 km) of extra travel per year—a lot of time spent walking instead of working.

Thus, for the cell-design team, arranging the cells to allow quick and easy access to all the parts, tools, and machines was key. In the linear arrangement shown in Figure 7-8, the two operators must travel from machine to machine in addition to moving the product from raw-material locations to work-in-process and then to finished-goods locations. Because demand patterns change, the design may need to include the ability to flex from two persons to three (or even four) and possibly down to one. An alternative to this linear layout is shown in Figure 7-9.

Because more people are right handed, and because right-handed people tend to work best from right to left, the consideration to design everything with a counterclockwise flow should not be overlooked, as shown in Figure 7-10. This is now a very

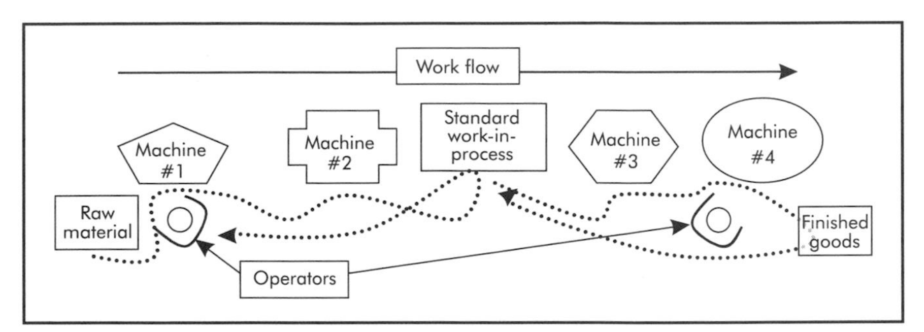

Figure 7-8. Typical linear work flow.

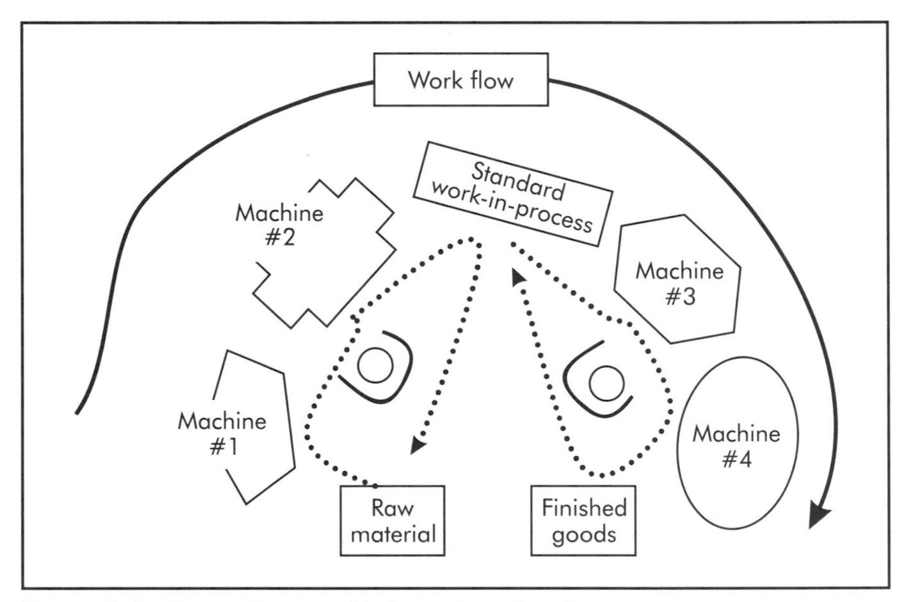

Figure 7-9. Conversion to a U-shaped work cell.

flexible operation capable of being staffed at levels to meet required demand.

Granted, the layout shown in Figure 7-10 may not look as neat and organized as if every machine was laser leveled and placed exactly the same distance from the wall. However, the goal is to be productive; geometric appearance is not a concern. Safety and quality are of primary concern along with effective manufacturing of product. When customers or other visitors tour the plant,

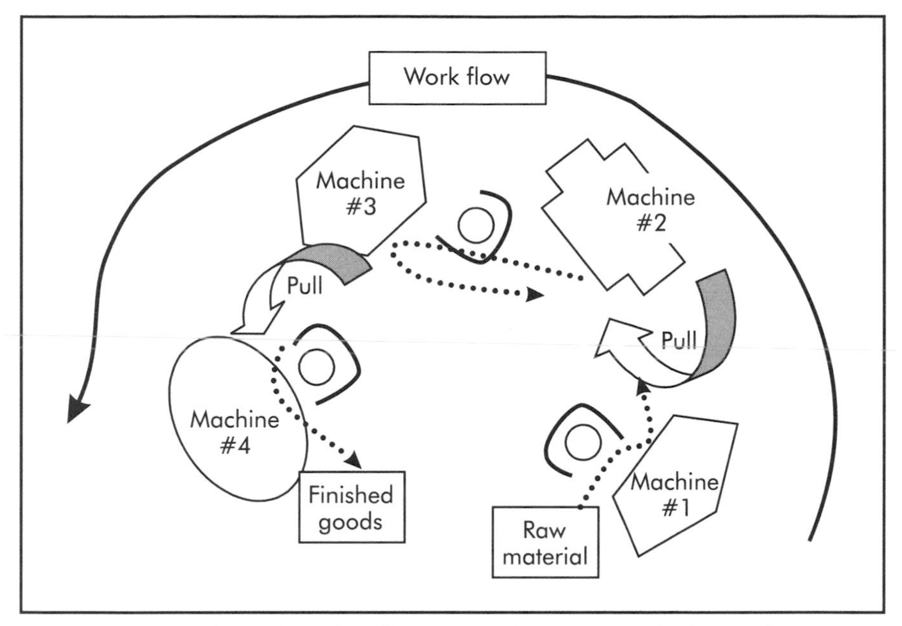

Figure 7-10. U-shaped work cell converted to counterclockwise flow.

they may be puzzled at the seemingly disorganized fashion in which the team has set up the cell. Thus, a little explanation will be needed to have these outsiders understand why the layout is preferred over more traditional configurations. Table 7-11 summarizes cellular manufacturing.

HIGH-VELOCITY MANUFACTURING

Everyone wants more speed. Customers demand speed and on-time delivery. A company probably makes similar demands of its vendors. We are all becoming increasingly impatient consumers of products and services.

It is not unusual to see people getting impatient, intolerant, or argumentative over the slightest inconvenience today—being made to wait for what they consider an "unreasonable" length of time. Thus, the company that can deliver the product or service first often has the best chance to earn the business. Having to say, "it's on back order" will normally kill a sale these days.

Table 7-11. Cellular manufacturing summary

Relationship to Six Sigma	Makes individual's work lives more meaningful and participative
Who needs and uses it	Value-stream managers, black belts, cellularization project teams
Cost	Moderate to high
Strengths	Gets people working together toward a common goal
Limitations	Can limit flexibility if designed around a few product types
Process complexity	Medium
Implementation time	1–6 months
Additional resources	See Bibliography
Internet search key words	Cellular work teams, cellular manufacturing
Internet URLs	www.informs.org/Conf/NO95/TALKS/WC10.html www.tcawcm.com/case2.htm www.tcawcm.com/Case1.htm

It's all about velocity. Speed of service or speed of product delivery is fast becoming one of the most important decision drivers for determining from whom a product or service is purchased. Of course, quality and price are also considerations. But if a company's quality is comparable and its price is within reason, a customer will give up a lot to get what is needed when it is needed.

Thus, the challenge is to never give customers a reason to go looking elsewhere. The biggest argument heard about reducing inventory and speeding up the flow of product is, "We never know what we are going to make. We have to group jobs together. Otherwise, we cannot plan week to week. We have to keep the pipeline full!"

Okay, so how does a company determine how many employees are needed? How much raw material is needed? The answer is usually, "Well, we know approximately what we will do in sales." Very well, so if it is known that $100,000 of a product will be sold next month, why is more than $25,000 worth of it in the pipeline if only one week's supply is needed at a time?

Regardless of what the final order file looks like, if the approximate sales volume is known, why not set up a process that allows that volume to fly through? Somehow, something will show up to fill the pipeline again. It always does. Or else a company wouldn't be able to forecast how many people it needs or how much raw material to buy. If there ends up being unexpected slack time because the expected order did not come in, why not use that time to perform cross training, preventive maintenance, 5-S activities, Kaizen events, etc.?

If sales orders are held and then released for batch manufacturing in an effort to smooth out the production schedule, time is spent anyway, moving parts unnecessarily, counting, inspecting, and so forth. It is just a hidden form of waste now. Consequently, there is less time for teams to participate in improvement programs because team members are spending their time performing non-value-added activities.

The reason that companies feel compelled to buy material requirements planning (or manufacturing resource procurement) (MRP) and tracking software modules is that traditionally it has taken "forever" for a job to meander its way through the shop, and when a customer calls to find out the status of a job, somebody better be able to provide an answer about its progress.

With high-velocity manufacturing, the job is either in a queue waiting to be released, or it is in a very short lead-time manufacturing process. The job does not need to be tracked if it's going to be done in a matter of hours. Only jobs that languish for days or weeks need to be tracked.

Inventory Reduction

"The more inventory a company has, the less likely they will have what they need." So said Taiichi Ohno. How true these words are, and yet how easy it is to be lured into the trap that many companies find themselves in today with huge inventories. Inventory has its place. It is the same as an insurance policy—a hedge against an accident or something going wrong.

You would never go out and get two, three, or even four different policies on your car and pay multiple premiums for coverage that you do not need. Yet, this seems to be the condition at some companies when you look at their inventory levels. They have two, three, or four times what they need and yet, just as Ohno said, they never seem to have the right thing.

Inventory reduction can create a wealth of opportunity while at the same time increasing the risk of dissatisfying a customer. What is the right balance? How much insurance is needed?

One company discovered that 12- to 14-week lead times were not uncommon. By Value Stream Mapping the process, finding out the true operator cycle time, and balancing work among the right number of employees, they were able to reduce the lead time in two departments from six weeks to two weeks. By doing so, the teams reduced the inventory by 33% (12 weeks reduced to eight). Their next goal is to further reduce lead time until they reach six weeks.

What is the motivation? Each month this company ships approximately $900,000 of this particular product. For each week of work-in-process inventory on the shop floor, it has tied up about $150,000 (or $600,000 any given month). By reducing inventory by four weeks, it will free up more than half a million dollars.

In addition, the company no longer has to ship a large percentage of products by air to maintain on-time delivery. Getting it done well within the customers' required lead times avoids shipping everything at $1.20/lb ($0.54/kg) and this translates to a $750,000 cost avoidance every year.

A recent article on the "DNA" of the Toyota Production System pointed out that while Toyota seeks to reduce inventory, they do not have a "zero-inventory" system. Toyota uses inventory sparingly, but when they use it, it is with a purposeful approach. They do not hide the inventory out of the way on a shelf or in a box. They try to place it in the way. If the inventory is hidden, there is no motivation to fix the problem that is creating the inventory in the first place.

Some inventory can be justified for production smoothing where there is a cyclical nature to the business. Generally speaking, inventory is a false god. It cannot save a company. Speed is the answer. If a company can process material fast enough it will have less inventory.

High-velocity manufacturing is summarized in Table 7-12.

Table 7-12. High-velocity manufacturing summary

Relationship to Six Sigma	Helps remove the lag times between processes, making it possible to see causes of bottlenecks and problems previously hidden by inventory
Who needs and uses it	Black belts, value-stream managers
Cost	Moderate
Strengths	Removes piles of costly inventory and is both an outcome and support to many of the other tools in the Six Sigma or Lean approach
Limitations	None
Process complexity	Medium
Implementation time	1–6 months (per value stream)
Additional resources	See Bibliography
Internet search key words	Manufacturing improvement, high-velocity manufacturing
Internet URLs	goalsys.com/id59.htm www.buker.com/

AUTOMATION

Automation has its place, but with smaller and smaller lot sizes, there is generally a long set-up component to the automatic device that counteracts the value of its operational speed or ability to operate unattended. Thus, automation may work for long-running jobs, but this does not represent reality in a job shop.

The test of whether automation is the right choice is to ask, "Will it improve the performance of the operator? Will it reduce the cost of the operator? Will it allow the operator to perform more value-added activities?" Sometimes, the automatic device simply ends up being watched by the operator. How has that saved any money or enriched the life of an operator who now has the mindless task of watching a robot eight hours a day?

The strength of automatic machines and robots is their ability to repeat the same motion without variation. If the quality of the process depends on minimizing variation, then there may be justification to automate it. Yet, robots and machines cannot reason like a human, so where there is a need to make a judgment as to surface condition, appearance, etc., the mechanical device can be automatically generating defects or passing them along to a human who must step in and stop the process and make corrections.

When robots are developed for specific tasks, the system's design engineer must examine the best set-up for the robot to perform effectively. It is interesting that when setting up work cells for people, this step is often overlooked. If the same checklist that a robot engineer might use is used to optimize an operator's performance, a company may find that it has less need for a robot!

Here are some of the issues applicable to robots (or people):

- Reduce the distance that the robot (operator) has to reach.
- Reduce the distance the part flows.
- Place all tooling and parts in the same exact location every time.
- Design racks, fixtures, containers and packaging so the robot (operator) can easily retrieve, manipulate, and discharge parts.
- Develop Poka-Yoke devices that make it simple to check part quality.

Another problem with automation is the dependability factor. For example, one company developed a new saw for automatically cutting metal forgings. Nearly four months after installation, the machine was still in a breakdown mode more often that it was operational. Understandably, the debug period can be substantial when developing any new tool, robot, or custom machine. There is no doubt the bugs will be worked out and this new machine will operate reliably soon. Nevertheless, in a cellular manufacturing environment, an unreliable machine is a great liability when the entire cell is affected by its breakdowns.

On the other hand, some automation is clearly a step upward. A simple example is bar-coding equipment where the tendency to create a typographical error is high for a human and rarely accepted by the computer readers. The effect of automation on Six Sigma efforts is covered in Table 7-13.

Table 7-13. Automation summary

Relationship to Six Sigma	Allows people more time to do the thinking work
Who needs and uses it	Industrial engineers, manufacturing engineers, black belts, project teams
Cost	Moderate to very high, depending on the application
Strengths	Frees operator up to do more meaningful tasks
Limitations	Automation is expensive, challenging to set up and debug, and may require significant maintenance.
Process complexity	High
Implementation time	3–18 months
Additional resources	See Bibliography
Internet search key words	Manufacturing automation, automation, robotic manufacturing systems
Internet URLs	condor.stcloudstate.edu/~amace/ www.managingautomation.com rd.business.com/index.asp

TEAM DEVELOPMENT

To be effective, a team must have a specific vision and shared goals. As a team matures, its members often develop new ways to communicate with each other. Watch a team operate for a period and you will see non-audible communication begin to develop. The more mature the team, the more invisible the communication becomes. A simple nod of the head, and people receive the message they need. A simple gesture takes the place of a dozen words.

Team maturity does not just happen; it grows through group dynamics as two or more people operate as one unit. However, giving teams too much responsibility too early can often result in frustration not only for the team, but for the managers of the value stream as well.

There are similarities between new teams and teenagers. Like teenagers coming into adulthood with much to learn, the team may think it knows it all already and wants to race ahead faster than management is comfortable with allowing them that freedom. The team may want to perform activities for which its knowledge, wisdom, and experience are not yet fully developed.

However, a team cannot gain that needed experience without the opportunity to try. What's a parent or a manager to do? Set attainable, reasonable goals. Six Sigma is about measuring performance. Figure out the measurements that the team should be focusing on and maintaining those will tell the story about its rate of maturity—such as how it manages Takt time and linearity charts, whether or not quality improves by 50% per year, and if 5-S audits and preventive-maintenance duties are performed in a timely manner.

When the fundamentals are being done well, more responsibilities can be added—such as participating in hiring new team members, managing a tooling budget, or developing vacation schedules. The best indicator of a manager's qualifications is how quickly and consistently his or her teams reach full maturity.

Mentoring and Delegation

The goals behind mentoring and delegation are to:

- duplicate or multiply the manager's own success in others,
- gain commitment,
- tap an employee's full potential,
- motivate others,
- ensure understanding of goals and objectives,
- speed up progress,
- enhance interest,
- benefit from the synergy of multiple minds working on a problem,
- capitalize on the creative nature of others, and
- keep the goals fresh in the minds of others.

What makes a good coach? A good coach:

- challenges people to do their best,
- sets a good example,

- never divulges a confidence,
- explains the reasons for instructions and procedures,
- is objective about things,
- lets employees make their own decisions,
- does not seek the limelight,
- listens exceptionally well,
- doesn't put words into others' mouths,
- keeps the promises he or she makes,
- works as hard or harder than anyone else,
- takes pride in developing team managers (leaders),
- gives credit where credit is due,
- never says "I told you so" or "That won't work,"
- gives at least one second chance,
- uses language that is easy to understand,
- handles disagreements privately,
- makes hard work worth it,
- gets everybody involved, and
- sets attainable milestones.

Recognition Systems

Effective recognition and rewards can range from a simple public thank-you between a manager and an individual to large-scale corporate events. It is always a challenge to put together a cohesive approach to employee recognition. A coherent process needs to be developed to ensure the system is even-handed and actually encourages results, rather than adding a source of rivalry or grievances.

Individual recognition should be done individually, and team recognition with the entire team. The goal is to avoid the "star mentality." Some teams fail because one or two members are always trying to garner all the attention, thus minimizing the recognition due the team. The principle for each team member should be, "I cannot succeed without my team."

Here are some things to think about:

- Keep track of all recognition, from simple informal events to major awards.
- Develop a matrix-driven scoring system that makes it easy to recognize the value of ideas, behaviors, or achievements.

Recognize the team's support for the idea, not just its being the originating source.
- Avoid any delay so that recognition is done as close to "real time" as possible.
- Evaluate the types of recognition given to teams or individuals in the past to ensure that any new forms of rewards are fresh and continue to motivate.
- Provide a strong link between cellular team goals and larger organizational initiatives.
- Ensure visibility of the recognition (especially team recognition).

Capitalizing on Synergy

Synergy can be defined quite simply as 1 + 1 = greater than 2. Synergism comes from the catalytic effect and action of separate people who together have a greater total effect than the sum of the individual's efforts.

A team's synergy is the effect of combining effort and cooperation while working together toward a common interest. Synergy results only if the team members recognize any weakness in their own ability, personality, and experience and are willing to seek the assistance of someone who has strength in that characteristic.

The forward progress of any initiative depends on more than one person. Educators have discovered in recent years that people learn faster and better in groups than when left on their own. Shared experience and problem solving as a team also speeds the transformation.

Group learning and problem solving must be carefully facilitated to avoid having the stronger personalities take over. With effective coaching, driver personalities can be helped to learn how to draw upon the analytic, social, and expressive characteristics and abilities of others in the group.

Brainstorming

The complexity and power of the human mind is nearly beyond our ability to comprehend. That we can even comprehend the brain itself is powerful testimony to the design of this electrochemical-biological wonder. We are the only animal with evidence of a desire and capability to consider the long-term effects of our actions. Other animals may be physically more powerful than we are, but only humans can acquire knowledge through shared experience. Something else that separates us is our ability to easily communicate ideas, a power that is most evident in a brainstorming session.

Brainstorming is a fun and productive method to generate ideas quickly. The first rule of brainstorming should be that there are no rules and no bad ideas. Sure, there are plenty of half-baked ideas, but no bad ideas. What is a half-baked idea? Like a pancake cooked on only one side, sometimes an idea thrown out at a brainstorming session is not fully cooked (or even appetizing) until it is flipped over and more fully developed, possibly in a later phase of team study.

An example of what seemed like a silly idea is the story about a toothpaste company struggling to maintain sales in a very competitive market. The future of the company did not look good, and management was willing to listen to all ideas. One employee standing in the back of the lunchroom piped up with the suggestion, "Why don't you make the hole bigger." Everybody laughed. What a half-baked idea! Yet, after they had thought about it, they realized that if the hole were bigger, people would squeeze out more than they needed each time and would end up using more of their product. So they tried it, doubling the size of the hole in the end of the tube, and toothpaste sales soon doubled.

Qualifying an idea (measuring its value) immediately during the brainstorming session will short-circuit the creativity of the participants, so the team should know that only clarifying questions are allowed. Here are some suggestions and guidelines for facilitating a brainstorming session.

- Clearly and concisely define the problem or opportunity.
- If the problem is complex, break it down into smaller problems.
- Make sure the team knows clearly the issues and problems involved.
- Pose an open-ended question to the group.
- Use the "round-robin" method, in which each person offers one idea at a time.

- Write each comment on a flip chart using the person's actual words.
- Do not judge the idea or allow others to judge it initially.
- Ask for clarifications if needed.
- Discuss and evaluate each idea.
- Have each team member anonymously vote for the top two or three ideas.
- Tally the votes.
- Ask the team if it can support the top selection.
- If 100% approval is not achieved, repeat the voting process.

Managing Meetings

Managers who call team meetings have an obligation to meet two primary needs of every participant: his or her personal and practical needs. Personal needs include:

- the need to feel valued and respected;
- the need to be listened to;
- the need to contribute to the discussion; and
- the need to share in the improvement.

Practical needs include:

- the need to use time effectively;
- the need to keep focused on the key topics;
- the need to reach a sound decision;
- the need to exchange information efficiently; and
- the need to feel like objectives were met.

Here are a few of the many reasons teams struggle with meetings:

- unclear objectives;
- unrealistic (too large) agenda;
- key people missing;
- apathy;
- lack of time management;
- alligators and rabbits (people who are all mouth or all ears);
- new group members (poor group dynamics);
- topics are uncomfortable to participants;
- inexperienced members; and
- short notice about the topic, little time to prepare.

To avoid problems like this, here are a few prevention techniques to try:

- Create an agenda with a clearly stated goal.
- Get the right people in the meeting.
- Use an issues board to capture non-agenda items.
- Educate participants about managing "air" time.
- Encourage people to bring data to support their position.
- Ask for help from the team when creating the agenda.
- Discuss time limits for each agenda item.
- Clearly state the benefits of a positive outcome.
- Measure the results.

Here are some intervention techniques to use if a meeting is going badly:

- Summarize topics as they are completed so people don't backtrack.
- Define the scope of meeting ahead of time.
- Address issues by writing them down for later discussion.
- Ask people to help keep the meeting on track.
- Declare that everyone participate and allow others to participate.
- Use of the procedural suggestion, "If I could make a suggestion."
- Refer to the agenda often during the meeting.
- Process check by asking, "What's holding us back?"

Fostering Change

The process of learning can be amazing. It results when someone or something creates a stimulus, the learner responds to it, interprets it, and then acts upon the stimulus by adding to it, relating, or applying it to something that he or she already knows. To get better requires learning. Sports figures know this very well: they study themselves, their competition, and then apply what they find out to what they already know.

A key aspect of learning is feedback. How quickly a company progresses toward its goal of being World Class will depend on how effectively it uses this learning tool. The power of the Six-Sigma approach is its ability to offer feedback in the form of visual and immediate information (stimuli), which the company can then act upon. Members of a Six-Sigma learning organization value the feedback of others and what data can tell them about the relative health (or lack thereof) of processes and ultimately the organization.

Although most people will say they do not mind change, the truth is, they only like change when it is their idea or if the idea makes their life better immediately (like a pay raise).

How can a company help people deal with change more effectively? If not careful, a company can lose good people during periods of significant change. It is important to recognize the resistance to change that can manifest itself in many ways. Here are a few indicators to look for:

- excuses, like "too busy,"
- delays in implementing change,
- avoiding training,
- absenteeism,
- drops in production, and
- sabotage.

Here are some techniques that can make the introduction of change easier:

- Establish a vision and clarify desired outcomes.
- Clarify roles and expectations from support groups.
- Describe the alternatives, including doing nothing.
- Verbalize the pros and cons of each alternative.
- Determine where to begin (where there is pain, where there is receptivity).
- Conduct briefings often (ask for patience, build networks).
- Get everyone involved (friends and foes in the same room).
- Recognize incremental milestones (celebrate small successes).
- Celebrate the process of getting there, not just the final achievement.
- Recognize that there will always be cynics and skeptics.
- Anticipate a vested interest in keeping things the same.
- Realize that not everybody has your assumptions or perception.
- Give resisters a voice but not control.
- Learn from failures (see breakdowns as opportunities).
- Maintain appropriate attitudes toward people (learn from the resisters).

Using New Technology

One of the greatest challenges that a team faces is limited resources. Teams need information. It costs money to staff a full-time training position, but there are alternatives for recording and re-transmitting key information. Videotapes, digital photos, Internet websites, and communication with FM radios are ways to capitalize on new technology to reduce the cost of supporting a meaningful training program.

If knowledgeable personnel have the ability to communicate with people who need that knowledge by way of radios, intercoms, or e-mail, then information and training has a better chance of being transmitted and understood.

Digital video and photos have been used to document complex set-ups and maintenance steps to show new operators or operators on other shifts how to set-up, disassemble, or maintain a key piece of equipment. As more computer terminals are put onto the shop floor, slide shows, drawings, and training videos can demonstrate to everyone the proper and improper use of tools.

Rather than throw or give them away, one company placed their obsolete office computers in the lunchroom. They were fine for installing math-blaster or blueprint-reading software so employees could sharpen their math or print-reading skills. Another benefit was that people became more comfortable using a keyboard and mouse even while playing games during lunch.

Table 7-14 summarizes team development.

CORRECTIVE ACTION

There are five steps to effective corrective action. However, most people only take advantage of just the first few, resulting in the problem coming back to haunt them time after time. If the entire package of corrective-action tools is used, then the issue that caused the problem should go away for good.

Here is a problem and the kind of questions used in an effective corrective-action loop. Through random sampling, a press brake operator discovers that he has been sending bad parts to the customer. For no apparent reason, the back gage (which sets the dimension) has a random and unpredictable problem and is not

Table 7-14. Team development summary

Relationship to Six Sigma	Changing culture is hard work, and Six Sigma must deal with all the elements of the organization. Teams are key to Six Sigma's ability to reduce the variability that results in poor quality and performance.
Who needs and uses it	Everyone
Cost	Low investment in capital, but high investment in time
Strengths	Sustained improvement is only possible by having employees who are fully engaged.
Limitations	Does not address the hard skills like set-up reduction, pull systems, etc.
Process complexity	Medium and made easier through the use of trained facilitators
Implementation time	1–5 years
Additional resources	See Bibliography
Internet search key words	Team development, team building
Internet URLs	www.teambuildinginc.com/index.html www.ravereviews.net/index.htm www.wm.edu/TTAC/ www.goalqpc.com

repeating as it should. This is the only machine of its kind in the shop and the only one capable of bending these parts, so until repair is made, bad parts are likely to be generated. Here are the five steps to correcting the problem.

1. Immediate response:
   - What do we do right now, since we've identified this problem?
   - Do we continue running the machine and sort out the bad parts?
   - Should the bad parts be thrown out or can they be fixed?
   - Do we sort through everything in the pipeline that could be suspect?

- Can an extra person be put over here to measure 100% of the parts?

2. Short-term response:

    - So the problem was found a little later than we would have liked and the customer may have received some bad parts, but what are we going to do about it?
    - Who will notify the customer?
    - Should the machine be shut down regardless of the effect on shipments?
    - Should everything shipped to this point be recalled and inspected?

3. Interim response:

    - The problem cannot be resolved until a machine-tool repair expert can get here, so what can be done until then?
    - Can a dial indicator on the machine be installed to tell the operator if the back gage has missed its mark?
    - Can the company farm out (contract) with someone else until this machine is repaired?

4. Long-term response:

    - Should the company invest in a new machine?
    - Should the company fix this machine or upgrade the back gage?
    - Would a preventive maintenance program avoid this problem in the future?

5. Root-cause analysis:

    - Why did the machine develop this problem?
    - Apply the technique of the Five Whys.
    - Is something inherent in this process lending itself to more problems like this?
    - What is the risk of this happening again?
    - Use an Ishikawa diagram (environment, method, materials, people, etc.).

The best solutions in the world are only as beneficial as the company's ability to implement them. So, be a good finisher. Close the corrective-action loops so the problem does not have a hope of repeating or resurfacing.

Corrective action is covered in Table 7-15.

Table 7-15. Corrective action summary

Relationship to Six Sigma	To ever achieve a 3.4 defects-per-million-opportunities level of performance, a company must have a meaningful closed-loop corrective-action plan.
Who needs and uses it	Everyone
Cost	Low to moderate, depending on the problem
Strengths	The five-step process ensures that very little is overlooked.
Limitations	Few limitations when used in concert with statistical-analysis tools
Process complexity	Medium, depending on the problem
Implementation time	1 day–3 years, depending on complexity
Additional resources	See Bibliography
Internet search key words	Quality programs, corrective action
Internet URLs	www.qualitysys.com/QMar.html www.assurx.com/company.html www.stochos.com/quality_action_reporting.htm www.epa.gov/seahome/cap.html

CONCLUSION

Upon completion of the Improve step, the team should be able to demonstrate the solutions that have been tested, simulated, or implemented. A deliverable from this phase is to show how each alternative scored in terms of cost, ease of implementation, resulting benefits, potential yield improvements, etc.

Team recommendations relating to implementation plans should be ready to deliver to the steering team. Short narratives should be provided explaining necessary changes to facilities, policies, procedures, equipment, materials, people, suppliers, transportation, etc.

A project summary outlining how each solution addresses the current condition expressed in the Define phase should be developed prior to meeting with the steering team or sponsor. Reinforcement of the measurement tools currently in place or needing to be established should be one of the deliverables of this step.

REFERENCE

Hirano, Hiroyuki. 1996. *5S for Operators: 5 Pillars of the Visual Workplace*. Portland, OR: Productivity Press.

8

Step Five: Control

Imagine driving down the highway without a center line or fog lines on the road. Although these lines are not the same as concrete barriers, they still help keep traffic from drifting onto the wrong side of the road or into the ditch. Tools like Takt time (manufacturing pace), line balancing, standard work, and standard work-in-process (WIP) inventory, used along with techniques like level selling and production smoothing, can all help increase product velocity. This chapter discusses how control mechanisms like these can help a company achieve the goal of Six-Sigma performance in terms of delivery.

Poka-Yoke, Jidoka, and Andon tools will show how the use of mistake proofing and visual signals can help operators identify when something has changed or is drifting toward a significant reason for needing attention. The use of best practices and the importance of process control are also discussed in regard to ensuring consistent, capable, repeatable, and reliable processes.

TAKT TIME

"Takt" is a term of Germanic origin meaning rhythm, and in the manufacturing sense, *Takt time* is the rate or rhythm of production. Major auto manufacturers have done a lot in developing improvement opportunities. However, calculating a Takt time is far more complex as the process evolves from a pure assembly operation to establishing a meaningful rate of production for a job shop.

In a job shop, to have a forecast (any forecast) for a known product and find the hours or minutes available for production is very different from operating in a world where everything is a mixed model. This is where it is hard to predict an order file or make any forecast. Added to this is the challenge that order quantities or due dates may change two or three times before the order ships or that no two products run at exactly the same rate. This is a recipe for stress-induced headaches and heartburn. Is this a good reason to disregard the use of this powerful tool? No!

Job shop management simply has to be creative to give people a meaningful objective (rhythm). The heartbeat of the manufacturing process is what keeps the company alive. If there is arrhythmia (erratic heartbeat) or no rhythm within the process, then no one can effectively use Kanban, work-load balancing, standard work, standard WIP, set-up reduction, or any of the dozens of other tools available.

Calculating a Takt time is easy:

$$T_t = A_T \div D \quad (8\text{-}1)$$

where:

T_t = Takt time
A_T = available time
D = demand or consumption

In an eight-hour day, there are 480 minutes. If two 10-minute breaks are taken out, this leaves 460 minutes. If machines are set up an average of five times a day, and it takes an average of 10 minutes to perform the set-up, that leaves 410 minutes. If a person is only effective 90% of the time (because of breaks, interruptions, etc.), then the net available time each day is only 369 minutes out of 480 (77%).

If the average demand is for 1,000 units per day, then Takt time is:

$$369 \div 1{,}000 = 0.37 \text{ minutes per unit (or 22 seconds per unit)}$$

If the average demand is for 250 units per day, Takt time becomes:

$$369 \div 250 = 1.48 \text{ minutes per unit (or 89 seconds per unit)}$$

This may seem remarkably easy until you mix in all the other parts that run at different cycle times and are ordered in various quantities. In this case, the best you can hope for is to deal in averages.

For example, assume that a company has three major product types, and within each product type there are six or seven distinct part numbers. Table 8-1 shows numbers for the average monthly sales volume for each part type and number.

Table 8-1. Average monthly build quantities

	Monthly Quantities		
Item #	Part Type A	Part Type B	Part Type C
1	350	100	700
2	240	25	350
3	500	30	350
4	125	15	350
5	70	60	
6	500	25	
7	25		

There are a total of 3,815 individual parts being ordered during an average month. Does Takt time work here? Remember, these are the order patterns for one month, so the numbers must be divided by 20 average days per month (Table 8-2).

Table 8-2. Average daily build quantities

	Daily Quantities		
	Part Type A	Part Type B	Part Type C
1	17.5	5.0	35.0
2	12.0	1.3	17.5
3	25.0	1.5	17.5
4	6.3	0.8	17.5
5	3.5	3.0	
6	25.0	1.3	
7	1.3		

Adding up all the average daily part quantities results in 191 units per day, so the Takt time calculation is:

$$369 \div 191 = 1.93 \text{ min per unit (or 116 seconds per unit)}$$

If a company were making one unit at a time, this would be the average rate at which each unit would need to be produced throughout the day. However, a job shop does not make exactly the same amount of every part every day. A job shop can work according to a hybrid Takt time based on the logical assumption that something will be made everyday, and on the average, it will look like this (although not exactly).

To set and maintain a uniform manufacturing pace, the operation must be able to distribute the work among all the workers at the same rate (116 seconds per unit). If one process takes twice as long (232 seconds), there will be a need to double up operators or run to fill a temporary standard WIP location to smooth out any variation in run times. This can be done on an off-shift until capacity can be increased, or until the team finds some other improvements.

Teams should use linearity (hour-by-hour) charts. These are not meant to be used punitively or as a quota system. When a team is dealing with mixed models and working in averages, some parts will run faster than others and the hour-by-hour chart for any one-day period may not always be perfectly linear. Over time, though, the effect of averages should show up as fairly linear weeks, months, and quarters. Figure 8-1 plots actual unit output against the linear planned goal.

Takt time needs to be adjustable in case of sustained changes in the order pattern. Minor fluctuations in demand patterns can be managed by a few extra hours per week, or a few extra days per month. Changing Takt time every week is not recommended. If people cannot sense the pace or maintain the pace for an extended period, there will always be arrhythmia in the circulatory system (the system that moves parts and makes money).

Table 8-3 summarizes Takt time.

LINE BALANCING

Back at the Rouge plant designed by Henry Ford, the distribution of work among just the right number of workers seemed a

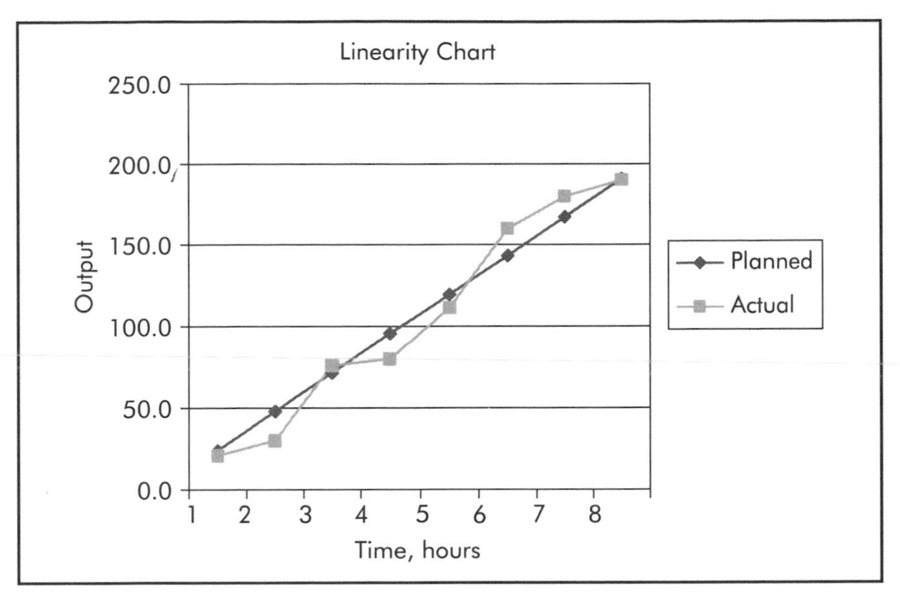

Figure 8-1. Chart of planned versus actual production.

Table 8-3. Takt time summary

Relationship to Six Sigma	Six Sigma is based on critical measurements and Takt time is a measurement that can be monitored, controlled, and improved.
Who needs and uses it	Product family leaders, value-stream leaders, black belts, green belts, cellular teams
Cost	Low
Strengths	Helps set the pace
Limitations	Does not focus on or impact quality
Process complexity	Medium
Implementation time	1–4 weeks
Additional resources	See Bibliography
Internet search key words	Takt time, rate of production, Lean Manufacturing
Internet URLs	isl-garnet.uah.edu/MEP/index.html www.atnlean.com/services/takt.asp

breakthrough idea. Generalists became specialists. In an ideal world, the manufacturing assembly line is perfectly balanced. If the goal is to produce one unit every two minutes, then everyone in the process has exactly two minutes worth of work to do. If one step in that process takes four minutes, two people are put on line doing the same task so that every two minutes, one assembled unit is ready. The trouble with this for a job shop is that each task takes a different amount of time.

For example, in the precision sheet-metal business, being a press-brake operator is as far as a floor person can go in terms of technical demands and need for special skills in blueprint reading, problem solving, math skills, and creativity. After attaining this status, it can be difficult doing the mundane tasks required with cellularization. New behaviors have to be learned to do whatever is needed and a team member must be willing to do whatever teammates require. In an auto-assembly plant, the work can be easily divided and shared among all in the assembly team. Each car or truck requires approximately the same amount of work, and the distribution of labor can be fairly stable, vehicle to vehicle. However, in a job shop, some parts may take much longer than others, so there are imbalances between workstations. In a sheet-metal plant, the first steps usually include punching the part out on a turret punch. Depending on the complexity of the part, this process could take from a few seconds to 15 minutes per blank. The next major steps generally include deburring and forming (bending) the part. A wide range of operator cycle times make it a challenge to balance the work between stations because what may punch in seconds could take minutes to form, and vice-versa. This is where the application of work sharing, and standard WIP comes in.

In a traditional sheet-metal shop, the material passes from department to department. The parts are punched, go into a deburring area, and then pass through a forming department and on to plating, assembly, or hardware departments. There are people assigned to do nothing but those activities, much like in Figure 8-2 where everyone does their own job and tries to stay busy doing just that all the time.

In a new Lean layout, where there is a need to balance workload, it is common to see operators overlapping to manage any

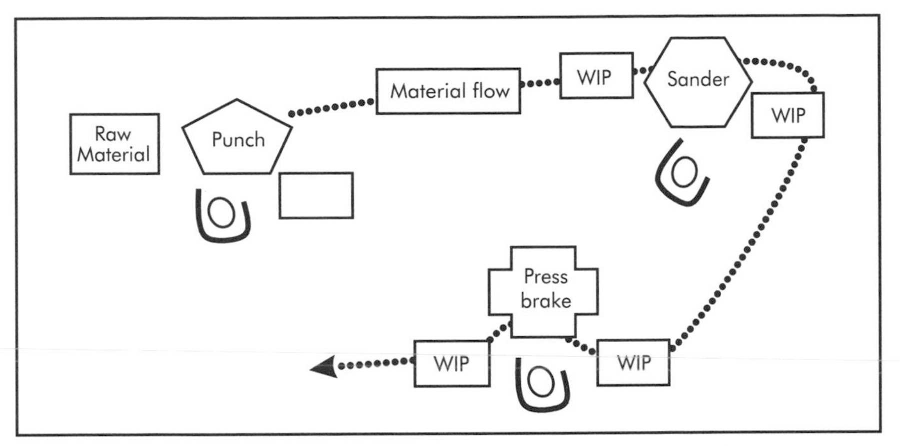

Figure 8-2. Typical layout with standard work-in-process (WIP).

imbalances in operator cycle time (Figure 8-3). Sometimes the punch operator has more to do than the press-brake operator, so the press-brake operator comes over to deburr the part. When the

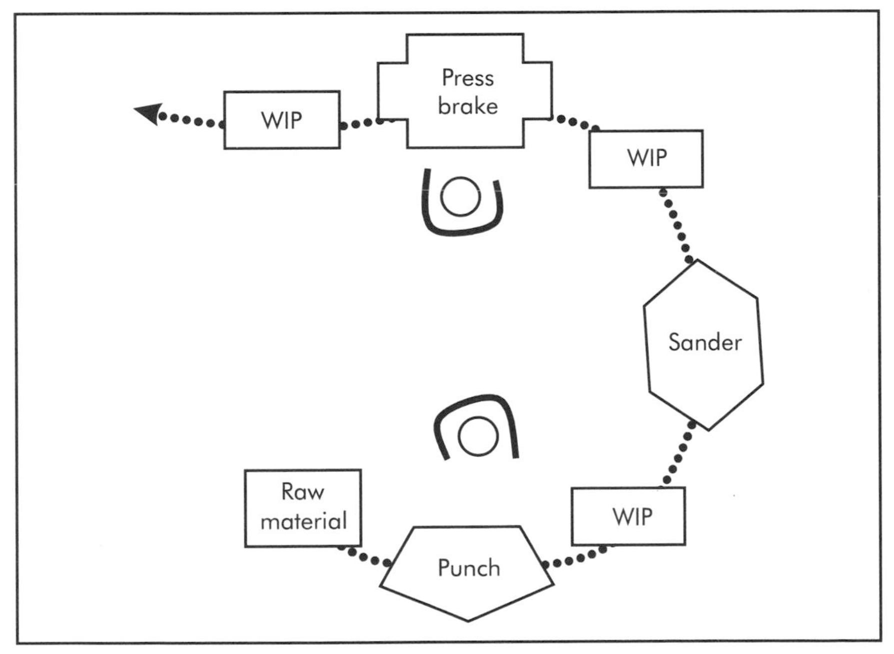

Figure 8-3. Reduced standard WIP with a counterclockwise flow.

punch operator gets ahead of the press brake, then he or she performs the deburring operation. This has the added benefit of being able to take out excess inventory. The work-sharing approach requires that each operator not only be trained but be willing to perform alternative tasks when needed. Standard work sheets can help identify the steps to take when there is a temporary bottleneck somewhere in the process.

If one of the machines (press brake or punch) is a constraint, and there is no opportunity for an operator to leave his or her machine, then an alternative to work sharing would be to manage the imbalances with a small amount of WIP inventory. This will help smooth out any material surge caused by one machine overrunning another. Company management may resist an idea like this. Extra training is necessary, as well as cross training on whatever function acts as a cushion to balance the work.

Production Smoothing

The goal of production smoothing is to keep the total manufacturing volume as constant as possible by metering work to the manufacturing floor at a uniform production rate. Kanban systems can generally deal with fluctuations in demand of ±10%. Heijunka techniques help smooth demand across the planning horizon. Having the fabrication department supply a mixed-model assembly line can help steady the flow of component production.

Standard Work

For a team to operate at Takt time (a known pace), there is a need to identify the work steps for each person for each cycle. This is easier to do in a pure assembly environment, but it also can be done (with a little creativity) within a make-to-order shop.

During a race-car pit stop, there are certain tasks that need to be accomplished. Problems are caused by any breakdown in communication or someone not performing their assigned tasks. Making things visible and easily repeatable is what standard work is meant to do. Anyone should be able to pick up the standard work instructions and follow them. Even beginners should be able to perform the standard work within a few days, but this does require simplification. The standard work sheets may change when

Takt time changes. People may be withdrawn or added to the team. Fewer or additional work steps may need to be added or subtracted from each operator's daily tasks. Standard work sheets can take many forms. An example is shown in Figure 8-4.

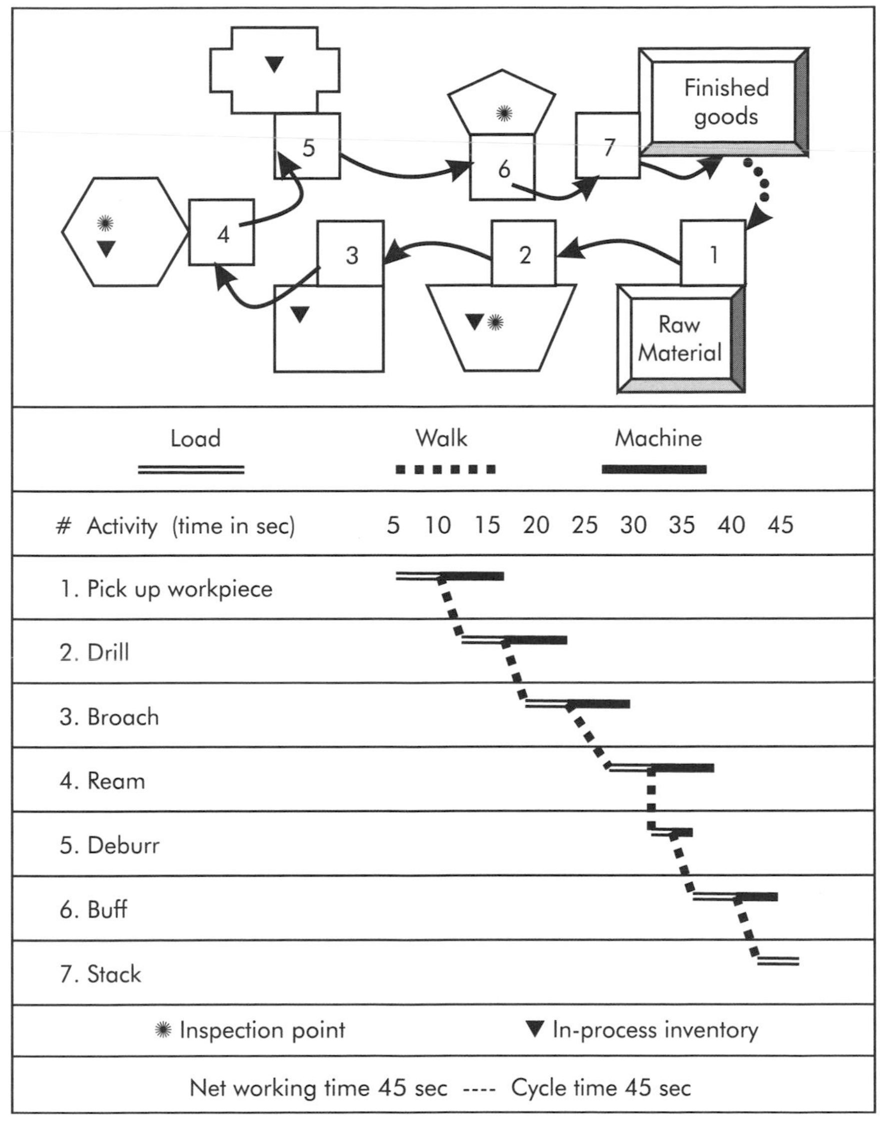

Figure 8-4. Sample standard work sheet.

Standard WIP

Standard work-in-process (WIP) has been mentioned numerous times throughout this text. It is a fundamental tool that can help smooth out minor imbalances in work flow, but it also can be a potential pitfall if not carefully managed.

The tendency is for standard WIP to grow and grow until it becomes hard to see material flowing at all. The term "standard" must be understood for standard WIP to work as it is designed. Standard WIP is a defined number of inventory pieces required to conduct the work sequence on demand. It is a known number of pieces, not a guess, not a fluctuating number, not a place that acts as a catch-all for overproduction, but the exact predetermined number of pieces allowed between two operations that will avoid starving either operation.

The discussion of line balancing emphasized the importance of work sharing to avoid the need for standard WIP. Where work sharing is not possible or until it is possible, standard WIP can help smooth out the flow.

Another technique for avoiding the use of standard WIP is to overlap the processes so that both team members are working on the same product at the same time rather than running all the parts of one job before transferring the parts to the next operator and then moving onto the next work order. Standard WIP has its place, but it should be used sparingly. If it is too easy to rely on, then there is less motivation to solve the problem that created the need for it.

Supermarkets

Your neighborhood supermarket works well to provide for your nutritional needs. If you had to go buy all your canned corn and green beans direct from the cannery, hamburger from the butcher shop, and milk products from the dairy, it would take many trips just to feed your family, and you would soon start buying in large quantities to avoid spending large amounts of time with all the scurrying back and forth. This is the same principle that leads to larger and larger batch sizes in organizations. It is simply too complex and takes too much time to order multiple shipments, and receive and transact multiple sets of paperwork if that process can be simplified by buying weekly stocks of material to feed the factory. Dealing with this issue is an essential goal of World-Class companies.

Having just the right amount of material without generating spoilage can be a delicate balancing act. The produce manager has a much harder time than the manager of the canned-foods aisle. Pure job shops obviously have a more difficult time with managing the need for seldom-used raw material than a captive house or pure-assembly operation where the same material is required on a routine basis.

Setting up small locations of material throughout the plant at the point of use rather than storing and retrieving the goods from a central store is much the same as the supermarket situation of saving the consumer those extra trips. Implementing Kanban cards on parts or their containers can become the signal between the factory (cannery) and the consumer (supermarket) that something has been used (consumed).

Level Selling

It would be nice if a job shop's sales had some degree of dependability: steady growth, unwavering, everything measurable and predictable. Realistic? Sure, if certain things remained constant. To make sales (or anything else) predictable and achieve a level of consistency, some key variables must be controlled. What are the causes and effects of sales? What are the inputs and outputs? What are the controllable variables?

If a company does nothing, it sells very little. Unless there is an established brand name that sells itself, a company must put some effort into marketing so that potential customers know what materials are available for sale.

Through the use and power of statistics, the likelihood and cost of getting each customer can be calculated. It can be determined how many quotations it takes, how many visits it takes to generate an invitation to submit a quotation, how many phone calls it takes to initiate a visit, and how many letters, advertisements, etc. To ensure that new sales (growth) and repeat sales are as level as possible requires an even distribution of sales effort all the time.

Assume there is a house-painting business. A crew of painters can paint one house per day, and the average house-painting job earns $1,000. It would be ideal to have four houses a week to paint as shown in Table 8-4. If it takes five quotes to get one painting job, then the estimator had better be quoting $20,000 in jobs per week. There is some referral or repeat business (add-on projects). So, this number can be reduced because one day a week is filled by word-of-mouth or repeat orders.

Table 8-4. Sales pipeline

→	→	→	→	Sales	←
Potential Customers	Phone Contacts	Visitation	Proposals or Quotations Written	New Projects or Sales	Add-on Projects or Repeat Sales
1,000/week	100/week	50/week	20/week	4/week	20%
$1,000,000	$100,000	$50,000	$20,000	$4,000	$1,000

Since five visits to potential customers are needed to generate an invitation to bid, $50,000 worth of potential painting customers per week (10 per day) need to be visited to generate the $4,000 in new business needed. To make 10 site visits per day, 20 phone contacts per day need to be made. This means $100,000 in potential project value in phone calls per week. To generate phone calls, the company must advertise. Of all the people looking for a house painter, there may be only one phone call for every 10 exposures to a newspaper ad. This equates to contacting $1 million of potential business to obtain $4,000 in contracts. Mailing and advertising must be done in a level fashion as well. If mailing or advertising is done all at one time, then all the follow-up phone calls will be batched as will the visitations, quotations, etc.

"Just-In-Time" selling leads to "Just-In-Time" buying. Granted, there is seasonality in many businesses (including house painting) and these cycles must be accounted for, but the message here is to keep the pipeline full at all times rather than having end-of-the-month specials or sales that create artificial spikes in demand and generate waste.

Here are some trends in forward-thinking supplier policies:

- Locate near the customer.
- Use small, side-loaded trucks and ship mixed loads.
- Establish small warehouses near the customer or consolidate warehouses with other suppliers (even competitors).
- Use standardized containers.
- Make deliveries according to a precise delivery schedule.
- Become a certified supplier.
- Accept payment at regular intervals rather than upon delivery.
- Manage stock for the customer.

What Six Sigma controls could be applied to your company's selling function?

Production smoothing is summarized in Table 8-5.

Table 8-5. Production smoothing summary

Relationship to Six Sigma	Uneven rates of production are hard to measure and improve. Balancing the work, assigning standard work, and avoiding the end-of-the-month push will reduce waste in many forms.
Who needs and uses it	Value-stream managers, black belts, green belts, cell teams
Cost	Moderate
Strengths	Helps support the application of Takt time and flushes out inventory
Limitations	Not a strict quality focus
Process complexity	Medium
Implementation time	1–3 months per value stream
Additional resources	See Bibliography
Internet search key words	Production smoothing, Heijunka
Internet URLs	fairmodel.econ.yale.edu/ www.utoledo.edu/~wdoll/LEANTHINKING/sld001.htm www.ame.org/main.php

MISTAKE PROOFING

We all learn from our mistakes, but what if many of our mistakes could be foreseen or prevented from happening in the first place? It is certainly a worthy effort for manufacturers, and the Japanese push for quality is credited with developing many useful techniques to eliminate mistakes and the waste they generate.

Poka-Yoke

Shigeo Shingo is attributed with the development of Poka-Yoke. The goal of this technique is to eliminate defects by preventing or correcting potential mistakes as early in the process as possible. Poka-Yoke is also called "mistake proofing" or "Murphy proofing" (Murphy's law states that: "if anything can go wrong it will"). It generally takes the form of a process design or device that makes it very hard for any person or action to generate a defect.

There are two types of Poka-Yoke devices: preventive and detective. A *prevention device* attempts to take away the opportunity for a mistake to be able to occur. *Detection devices* notify an operator that a fault has occurred.

For example, in an attempt to Poka-Yoke driver safety, automobile manufacturers a few years ago attempted to make it impossible for people to start their cars until all seat belts had been fastened. This was a good example of a prevention device. However, consumers revolted, and after receiving many complaints, the carmakers caved in and replaced the system with an annoying buzzer (a Poka-Yoke detection system). Even then, some people would disarm the buzzer so they would not have to hear the warning. Poka-Yoke devices are only useful if they are used. If they are obtrusive or intrusive, people may disarm them or ignore their use.

Some rules of thumb when developing Poka-Yoke devices are:

- make them easy to use (simple);
- make them inexpensive (creativity before capital);
- build them into the process;
- make them capable of measuring 100% of the output; and
- place them close to the process where mistakes are likely to occur.

Some examples of Poka-Yoke devices are:

- light curtains that disable a machine if someone walks through the light beam;
- unleaded-gas nozzles a different size than leaded-gas nozzles;
- Videotapes with a break-off tab to avoid recording over something you wish to save;
- computer floppy discs that install only the right way; and
- microwave ovens that will not operate with the door open.

Jidoka

Jidoka devices are mechanisms that stop a process the moment a defect or abnormal condition (potential defect) occurs. The faster that corrective action is taken, the greater the likelihood that defects will not be generated. Although no one wants to see a production line stop, it may be the best thing that can happen to avoid generating scrap.

For example, a team operating a wood-laminating machine was so focused on output numbers (board footage) that when a potential defect occurred, they would just keep running until the press load was complete to be sent to the oven to cook.

In observing the team for one day, it was noticed that the glue-application device was skipping or missing certain parts of the wood, resulting in delamination or separation at the glue line and the board had to be scrapped. The team did not stop for the one-minute it would take to clear the tube leading to the glue applicator because the normal oven cycle would be missed. Instead, team members would smear the glue by hand as well as they could, and hope that the parts would not delaminate. As this event was watched, 22 boards delaminated.

With a few calculations, this team was shown why their decision to risk a delamination (defect) was not a sound business decision.

- Eight people were on this team.
- Average wage was $12.00 per hour.
- Average value of a laminated board was $10.00.
- Average press load was 22 boards.
- Average time to clear the glue supply tube was one minute.

If all eight operators had to stop for one minute, it would cost the company $0.20 per person or a total of $1.60 in labor, but by risking 22 boards having to be thrown away, the company lost $220.

Okay, but then how could they make sure that the glue-application device was working properly at all times? A Jidoka device could measure the weight of the board before and after the glue application to ensure that the correct amount of glue is applied. If not, the machine shuts down.

Andon

An Andon, literally meaning "lantern" in the Japanese language, is an alarm lamp most often used to signal an abnormality has occurred. Andon lights can be used to signal the operator, other team members, material handlers, maintenance personnel, team leaders, supervisors, or department heads of an abnormal condition whenever it is appropriate for them to know about it. Andon systems can also include audible signals like buzzers or horns. The alarm system can be tied to a Jidoka device that stops the machine (itself a signal).

In precision sheet-metal shops, some laser-cutting and turret-punching machines are equipped to run "lights-out"(when no one is around), automatically loading and unloading sheets of material. If the machine accidentally loads two sheets, or if one of the sheets misloads, a signal is sent to a computer, which then places a phone call to the operator (even at home).

A visual tool also can be used for helping to keep teams focused on set-up downtime. For example, if the set-up goal of 10 minutes is approaching, a light could notify the team that they are behind schedule. Mistake proofing is summarized in Table 8-6.

PROCESS CONTROL

Imagine setting up and running a laser-cutting machine. There are many variables: wattage, feed rate, focal length, assist gas, gas pressure, material type, material thickness, material condition, humidity, ambient temperature, time of day, or even the opening of the shop door. To control the process, the operator has to eliminate or control any variables in the process. When faced with

Table 8-6. Mistake-proofing summary

Relationship to Six Sigma	Eliminating the potential for a defect is the best way to attain a Six-Sigma level of performance.
Who needs and uses it	Everybody can implement Poka-Yoke devices in their work area.
Cost	Low to moderate
Strengths	Fixes problems once and for all
Limitations	Some tendency to over-engineer the fix if teams are inexperienced
Process complexity	Medium
Implementation time	1 day–3 months
Additional resources	See Bibliography
Internet search key words	Poka-Yoke, Jidoka, mistake proofing
Internet URLs	www.fredharriman.com/services/glossary/tps.html www.dig.bris.ac.uk/teaching/m_o_i/studen10.htm

a problem, a process of elimination is used to find the cause of the problem and then settings or conditions are changed until the problem goes away.

By using techniques described throughout this book, some of the causes of process variables can be reduced or eliminated. Through the use of Design of Experiments and scatter diagrams, the best possible conditions can be isolated for the most optimum results. The power of statistical feedback is well worth the time and resources it takes to collect the data and learn to interpret what it tells about a process.

Best Practices

Without stable, repeatable processes, consistent, dependable output cannot be attained. Although creativity should not be stifled or the potential improvement overlooked, there needs to be a formula for this "experimenting." There needs to be control over what and when it is acceptable to test new ideas.

A business needs to ensure stability in the quality of its products or services, and consistency in what the customer sees as a final product. This demands controls in place to make it a "best practice" organization. Some organizations call these controls standard operating procedures (SOP). They are little more than a page or two that describe the minimum steps required to ensure that everyone follows the same process. Checklists, text, photographs, videotapes, computerized work instructions, samples, protective light curtains, and other Poka-Yoke devices can all be used to ensure that an adequate understanding of the best practice for a process is understood and shared by each operator. Process control is summarized in Table 8-7.

Table 8-7. Process control summary

Relationship to Six Sigma	The "C" in the Six-Sigma DMAIC cycle is all about control.
Who needs and uses it	Everyone who manages a process
Cost	Low to moderate
Strengths	Helps eliminate variation
Limitations	Does not focus on productivity or profitability issues
Process complexity	Medium
Implementation time	1–4 weeks per process
Additional resources	See Bibliography
Internet search key words	Process control, quality, best practices
Internet URLs	www.margaret.net/spc/ www.benchmarkingreports.com www.masterytech.com/bestpractices.htm

CONCLUSION

Evidence of the adequate completion of the Control step includes being able to show and explain to others the tools in place to standardize the process across all shifts, departments, machines, people, and divisions. There should be random audits to provide confidence that behaviors have actually been modified to align with new or approved procedures and best practices.

A formal presentation to everyone affected by the change is a good idea. It also provides a chance to reinforce the learning for the team that just went through the process. If there is a need for additional teams, this presentation becomes the "baton" that is handed off to the new team.

9

Other Tools

It is very rare to find a company that is totally self sufficient; one that does not rely on a supplier or anyone else for the quality of an end product. Therefore, supply-chain management is critical to every company. Of course, it may be best to bring your own process under control before seeking improvement within the balance of the supply chain. It should be obvious how critical suppliers are to achieving a goal of Six-Sigma performance.

Development of in-house champions, black belts, and green belts (important players in carrying out Six-Sigma projects) may take on a more formal approach in larger companies. However, the same basic principles and techniques are used by continuous-improvement teams in the smallest companies where the resident black belts may have many other hats (and tool belts) to wear.

SUPPLY-CHAIN MANAGEMENT

A *supply chain* is the network of raw-material providers, manufacturing, or service facilities and distribution systems that performs the function of material procurement, transforms materials into either intermediate or finished products (subcomponents), and finally distributes these products to customers. The entire network of companies works together to design, produce, deliver, and service products.

Since its advent, the field of supply-chain management (SCM) has become tremendously important to companies in an increasingly competitive and global marketplace. Shortened product life cycles, increased competition, and higher quality and lead-time expectations from customers have forced many companies to reexamine their physical logistic management approach in favor of a more holistic supply-chain management strategy. In the past, companies focused primarily on manufacturing and quality improvements within their four walls. Now, their efforts can extend far beyond those walls to encompass the entire supply chain.

Many companies on the path toward World-Class status have had to come to grips with the true meaning of supply-chain management. Some companies are ambitiously implementing SCM in both the internal and external portions of their supply chains. The levels of complexity of supply chains may vary greatly from company to company and industry to industry. Supply chains exist in both service and manufacturing organizations.

Recent and rapid developments in information and communications technologies have enabled the effective implementation of SCM. Creative use of databases, communication systems, and computer software has been crucial in the development of cost-effective SCM.

In simplest terms, the supply-chain loop starts with the final customer and ends with that customer. All information, transactions, materials, and finished products flow through the loop. SCM can be difficult. The supply chain is a complex network of facilities and organizations with dynamic and sometimes conflicting objectives.

In recent years, it has become clear that many World-Class companies have already reduced their manufacturing costs as much as is practical. Therefore, in many cases, the only possible way to further reduce costs and lead times is to look into the extended value stream that includes vendors. Thus, the mature, Lean company leads its suppliers toward removing waste from their processes through effective SCM techniques.

SCM can be compared to a well-balanced and practiced relay team. The team is more competitive when each member is properly positioned for the hand-off. Relationships between players who directly pass the baton to each other are the strongest, but the entire team needs to coordinate to win races. For the manufacturing "hand-offs" to work seamlessly, critical decisions have to be made regarding location, production, inventory, and transportation for the supply chain to operate well.

Location

Locating production facilities, stocking, and sourcing points is the first step in creating a supply chain. A key decision is the design of alternative material paths for product to flow through to the final customer. This is where some of the other tools in this book come into play.

In one case, a raw-material processing facility was located 2,200 miles from the first point of use because that is where the raw material grew (trees). To locate the lumber mill closer to the manufacturing and assembly facility would mean transporting 34% of all waste (chips and sawdust) half-way across the United States.

Other considerations for locating facilities deal with labor costs, taxes, duties, and tariffs.

Production

Obvious manufacturing decisions include what products to make and which teams will produce them. Production scheduling, development of master production schedules, scheduling machines, and equipment maintenance are all important tasks. Critical considerations include balancing operator cycle times (workload balancing) and taking quality-control measures during all stages of production.

Inventory

Inventories must be managed whether they consist of raw materials, in-process, or finished goods. The primary purpose of inventory is to smooth, insulate, or buffer production against uncertainties in the supply chain. Holding inventories is always costly, and thus efficient management is critical, especially because these levels determine the ability to meet customer demand.

Transportation

By reducing work-in-process (WIP) inventory, one company was able to reduce lead time, which in turn reduced the amount of money spent shipping everything by express mail. Since the best choice of transportation is often a trade-off between cost and time, there was an incentive to find methods to save time elsewhere. To do so, the company used tools like Kanban and inventory reduction described elsewhere in this book.

Air shipments may be fast and reliable, but they are expensive. Shipping by sea or rail may be cheaper, but large amounts of inventory must be held, and there is an inherent uncertainty about time. This brings us back again to location. A company's geographical location in relationship to its customers will impact all the transportation decisions within its supply chain.

Pooled loads (transportation companies moving dissimilar material to the same area) or shared truck space to the same customer (even between competitors) is becoming more commonplace.

Supply-chain management is summarized in Table 9-1.

CHAMPIONS, BLACK BELTS, AND GREEN BELTS

To train a fully competent black belt (dedicated continuous-improvement champion) may take months or years and can be

Table 9-1. Supply-chain management (SCM) summary

Relationship to Six Sigma	Everything is open to scrutiny when attempting to make major improvements, including the supply chain.
Who needs and uses it	Purchasing managers, black belts
Cost	Moderate
Strengths	Examines the material supply chain to find potential causes of defects, unwanted inventory, long lead times, and wasted motion or effort
Limitations	The company may be at the mercy of suppliers who are unwilling to participate.
Process complexity	Medium to difficult
Implementation time	1–5 years
Additional resources	See Bibliography
Internet search key words	Material supply chain, supply-chain management (SCM)
Internet URLs	www.supplychainseminars.com/ www.bpubs.com

expensive. This book will not describe every tool used in the life of a black belt. However, it will briefly explain the most common ones and particularly those that can be put to use right away, rather than after months (or years) of study in higher-order statistics and methodologies.

Although it may be questionable whether a small job shop can financially support a full-time team of black belts, the idea of someone (or a team) acting as green belts (part-time resources) should be a consideration for any company. The skills of these two classifications have less to do with natural ability, and more to do with experience and the amount of time dedicated to the role of facilitator or project leader. Involved in Kaizen (continuous-improvement) projects, these team leaders should be able to bounce in and out of projects as the company sees the opportunities and can rationalize the cost of conducting them. This is not to say that there should not be structure. Many companies have realized value in the old saying: "People do not plan to fail; they fail to plan." Therefore, a strategic plan and budget to support that plan should be adequately addressed.

Many companies have seen the value in setting aside budgets of 1–1.5% of sales for continuous-improvement projects. When projects are identified, teams can then make applications to fund Kaizen events and improvement ideas from this budget.

There are arguments that setting up master black belt, black belt, green belt, or any other color belt may foster a class society within the company and could be counterproductive. Although only time will tell if this is a problem for a small company, it is the author's opinion that the structure of the black belt or green belt program works best in large companies where the superstructure supports it, and where efforts might otherwise be easily hidden by the sheer size of the company itself.

Regardless of the title given a person, the goal is to get everyone involved. This is especially important in smaller organizations where resources need to be extremely well managed. The title of continuous-improvement manager or champion is really not that different than black belt or green belt, but whatever the nomenclature, a company should avoid developing an organizational class society.

Training

Many companies have found that having an internal resource for running Kaizen events and longer-term projects can be more economical than relying on outside resources such as consultants. This requires that the in-house resources be well trained in project management, team building, and use of Six-Sigma techniques and tools.

Some companies set up prerequisites for those desiring to become a black, brown, or green belt. They may have to complete a bachelor-of-science degree, a master's degree, or other certificate of completion in particular disciplines.

Effective black-belt and green-belt training programs combine traditional classroom-based teaching along with providing participants hands-on experience. Typical workshops lead candidates through a carefully structured, supervised series of study topics, discussions, and exercises over a one- to two-week period. This preparation is usually done by teleconferencing, distance learning, e-mail, or through reading assignments. These sessions prepare participants for the classroom training that follows. Six-Sigma tools are reviewed and statistical tools are studied in-depth. This step is generally done off-site (away from work or at a site-based training center) so the candidates are completely immersed in the experience.

The test of any effective training program is its ability to provide candidates with both a technical and practical foundation for improving business processes. Objectives when seeking a training provider include:

- Provide participants with a thorough understanding of Six-Sigma methodology.
- Impart the knowledge needed to apply the Six-Sigma tools as a team leader.
- Provide participants a working knowledge of the required statistical (DMAIC) tools.

A typical Six-Sigma course outline includes:

- a course overview;
- Six-Sigma overview;
- DMAIC methodology overview;

- Define step;
- project definition and charter;
- Measure step;
- process mapping;
- basic statistics and graphical methods;
- measurement systems;
- process capability;
- Analyze step;
- Failure Mode and Effects Analysis (FMEA);
- Improve step;
- Design of Experiments (DoE);
- Poka-Yoke and other improvement tools;
- Control step;
- Statistical Process Control (SPC); and
- process-control plans.

A sound training program should emphasize:

- applying Six Sigma and Lean business tools to specific company objectives;
- achieving measurable business results;
- mentoring other candidates;
- establishing a sustainable continuous-improvement program;
- strategies for problem solving;
- use of SPC;
- use of FMEA;
- Measurement System Analysis (MSA);
- designing a visual workplace;
- facility layout; and
- reducing set-up time.

Black belt training is summarized in Table 9-2.

Table 9-2. Black-belt training summary

Relationship to Six Sigma	The workhorses of a Six-Sigma program (whether called black belts or something else) need to understand the tools of the trade.
Who needs and uses it	Black belts, green belts, project leaders, team leaders
Cost	Low for informal training, moderate for formal
Strengths	Provides critical in-house capability so that a company does not have to rely on an outside resource.
Limitations	Takes time to train
Process complexity	Medium
Implementation time	1 year
Additional resources	See Bibliography
Internet search key words	Black belt, Six Sigma, quality champion
Internet URLs	www.i6sigma.com/library/content/c010618a.asp www.inel.gov/inews/current/05016sigma.shtml www.sme.org/cgi-bin/get-newsletter.pl?SIGMA&20020425&2&

10

The Light-bulb Factory

Every team member must keep in mind that a total transformation is not possible without full and complete adoption of the Six-Sigma vision. Everyone at all levels in the organization must see their role clearly. The analogy of the light bulb will be used to symbolize the moment when people recognize how their activities affect the organization's World-Class initiatives.

How many light bulbs does it take to change a company? Like many things, light is often taken for granted; little attention is given until it suddenly goes dark. Then, we realize our dependency on those little photon-generating devices.

A burned out light bulb is annoying and, in some cases, even deadly. No one likes being left in the dark. The light bulb symbolizes ideas—new and older burned-out ones—that relate to manufacturing processes and methodologies. Some ideas must be modified or replaced if a company is to move up to World-Class manufacturing methods.

The light-bulb analogy is also meant to highlight the moment in a person's life when he or she recognizes a completely new and stimulating concept. The common term for this experience is an *epiphany*, defined as: "A sudden, intuitive perception of or insight into reality, or the essential meaning of something often initiated by some simple, commonplace occurrence." This is the light-bulb moment. An organization must recognize the need to modify its past practices if it is to move forward toward a new future. So, how many epiphanies does it take?

THRIVING IN THE HEAT

Once, there was a man in the high desert, who had only limited success farming using the techniques he learned as a child. He harvested a few scrawny ears of corn and an occasional cherry tomato. It was frustrating. What the rabbits couldn't reach, the deer consumed. What the varmints did not strip clean of foliage, the morning frost and scorching afternoon sun destroyed. What worked in the moderate climate of one area clearly did not work here. To be successful, the farmer would have to adopt a radical approach by being re-educated on the techniques necessary for success in an arid environment. His neighbor, an older gentleman down the road had a beautiful, thriving farm. He was adapting by using special techniques.

The same is true in business. When the business environment changes, companies often can not use the same sales, planning, and manufacturing techniques that worked in the past. If companies continue to do things the old way and refuse to notice what more successful companies are doing differently, then they are doomed to watch their business "farms" flounder, get eaten away, or burned up in the scorching heat of competition. Being open-minded and willing to learn new approaches can be the difference between thriving and merely surviving.

THE NEXT NEW ECONOMY

In the past 200 years in America, there were five major economic generations in which some nimble entrepreneurs got rich as these new industries changed the world forever:

- steamship;
- railroad;
- electric motor;
- automobile; and
- communication (technology).

In their early stages, each provided significant economic opportunities and growth. There were also periods of significant economic downturns and financial losses for many less fortunate individuals and organizations when some reached their maturation or saturation phase.

There is never one magic bullet for surviving the inevitable economic downturn. The keys to survival are many, varied, and usually very specific to each situation. This book is meant to identify and explain potential survival techniques used successfully by World-Class organizations during past and present economic cycles. It is not about resurrecting dead technologies or trying to save them in the face of improved methods. It is about how to sustain and improve companies, and how to remain viable and competitive regardless of new technologies or new industries. To be the product or service provider of choice, a company requires flexibility, adaptability, and creativity.

For example, if a company were around during any of the previous industrial eras and owned a sheet-metal-fabrication firm, it could have provided materials to customers. Yet, to be a viable supplier—then or now—it would have to be competitive in three primary areas:

1. deliver on time,
2. meet quality standards, and
3. comply with cost requirements.

A customer who is looking for a supplier wants them to be competitive regardless of the era, geographic region, or economic generation. To win that business—a hundred years ago or today—a sheet-metal company must provide incentives for the customer to buy from it rather than a competitor while meeting all three of the fundamental expectations. Moreover, to keep that business year after year, a company must also maintain a reasonable profit margin. This is the goal behind the Six-Sigma approach: making a company a viable supplier of goods regardless of the economic phase of the moment or what lies ahead.

LEARN TO BE LEAN

When an economic slowdown occurs, it forces companies to become more concerned about how to encourage employees to seek improvements in productivity: all while the employee is seeing his or her co-workers sent home due to a softening economy. The application of World-Class techniques does not result in layoffs, yet the pain of seeing co-workers displaced (regardless of the cause) is hard to deal with. How can the vision of the Lean enterprise be championed within an environment and economy of eroding sales, and cutbacks in spending and staff?

Certainly, the economy presents many challenges, but also a unique set of opportunities. In the heat of battle, there is rarely time to spend on education and training, defining and documenting processes, analyzing and reducing machine set-up time, and other improvement activities. Yet, if the current economic situation is looked at as a chance to prepare for the next wave, a company can begin to distance itself from the competition. This does not mean buying new buildings or equipment, but rather investing in improving methods, including the human element. Companies can prepare to exhibit World-Class performance by examining and modifying the overall approach to running business, including such manufacturing fundamentals as set-up reduction.

Sports teams take time to watch videos of the competition and themselves to find areas where they can improve. Taking time to visit other companies who are also on the path toward adopting a World-Class approach is a beneficial activity for off-peak periods. In learning any new skill, the best way to learn has always been to watch someone else who knows what to do, or to learn with the help of a coach or trainer.

When trying to implement a new initiative or set of technologies like Six Sigma or Lean Manufacturing, attempting to adopt the new methods without coaching can be very challenging, especially if no one in the organization has experience with the new skills and behaviors. Lack of coaching can add significantly to the time required to adopt the new method.

Reciprocal visits with a company further up on the learning curve are a powerful method of demonstration and education from which everybody benefits. The learners get knowledge, and the teachers get to reinforce the learning for themselves. These reciprocal visits can be all the more powerful when the companies visited are also customers. They get to see your commitment to follow their lead, and you get a chance to talk with their front-line workers and build new relationships.

Here are just a few suggestions for activities that will pay huge dividends while companies wait for the economic faucet to get turned back on:

- set-up reduction,
- quality enhancements,

- machine reliability and capability gains,
- experimentation with a virtual cell,
- cross-training,
- workplace organization (5-S activities),
- teamwork training,
- standard work definition, and
- material-handling improvements.

Think for a moment what would happen if a company like Toyota decided to get into your business and opened a plant next door. Imagine also that it had a copy of your customer list, and began applying the same highly productive techniques it has applied to the manufacture of automobiles. Also assume its employees are twice as productive as yours, and its lead times are three weeks and yours remain at six weeks.

What would have to be done differently to compete with a company like that? This question must be asked to create the sense of crisis to motivate people and organizations to reach for the next level of performance. Make no mistake, even if this imaginary scenario has not happened yet, a company still needs to get ready because it will happen eventually. It may not be Toyota, but it will be a company that has adopted Toyota-like World-Class manufacturing capabilities.

If a company is a supplier, subcontractor, or make-to-order manufacturer, it is selling manufacturing capabilities, not computers or cars. Application of work force skills, equipment, and capability is limited only by flexibility, imagination, and the needs of the marketplace. If a company is a great supplier, it will no doubt be a great supplier regardless of what the market demands.

Henry Started It

We all know how Henry Ford's Rouge plant took the typical role of the automobile "artists" of his day to a completely new level. Ford removed the black magic associated with hand-sculpting body panels and components. He transformed the artful fabrication processes into more scientific and repeatable operations. His approach (an epiphany no doubt) was a radical transformation. It was so significant that even decades later, manufacturing and industrial engineers had a hard time envisioning improvements to his process of transforming raw materials (as basic as iron ore) into a finished automobile in just a few days.

In recent decades, companies like Toyota, Harley Davidson, GE, and John Deere have not sat idly by, accepting the Rouge plant as their model. They have honed their manufacturing processes by applying a holistic approach to improvement programs. By involving everyone at every level of the organization, Toyota has taken the auto industry to a new level.

Many organizations have wilted in the heat of new competition. They have not necessarily been defeated by Toyota, but by new thinking and new methods applied to old processes. In some cases, they have been beaten by their own unwillingness to recognize, adopt, and adapt to proven and improved methods. It could have been anybody who discovered the power of a waste-free manufacturing approach, it just happened to be Toyota.

Just as Toyota realized that the Rouge model needed to be modified, and GE understood that its quality system needed an overhaul to meet the needs of a more sophisticated consumer, so any company needs to realize that even the Toyota model may not hold all the answers as industry transitions into the next new economy. The light bulb must go on for all.

Everybody is a Light Bulb

To adapt and survive in today's new competitive environment, our old ways of doing things must be challenged, recognizing that how a company may have operated successfully in the past may no longer be a valid approach. Much like a chain reaction, that reaction must be sustained. A company will not serve its stakeholders, customers, employees, or even its community well if only one person is responsible for generating or sustaining all the enlightenment this requires. People come and go. The future of the company should not be put at risk because there is only one person to act as the catalyst for change.

There's an old light-bulb joke that goes like this:

> Question: How many country-western singers does it take to change a light bulb?

Answer: Two, one to change the bulb, and one to sing about how much they miss the old one.

There is no way to avoid the fact that change comes hard. It is harder for some people than others, but the worst thing imaginable is to spend time and resources making needed changes, see the change effectively implemented, and then have the changes come undone because of strong, change-resistant personalities. About 80% of the people in the world are protection motivated. That is, prone to resist change, and worse yet, prone to undo change if permitted to do so.

If there is a perception within a company that resisting change is okay, it will significantly decrease the chance that change will be sustainable. Protection-motivated individuals want things to stay just as they are. To avoid backsliding, a new commitment at every level is required. Two steps forward and one back are better than no steps forward. However, if the steps backward are due to individuals undermining or sabotaging the improvement efforts with little or no censure for doing so, then positive change will be hard to sustain. If a company has spent time and resources changing the figurative light bulbs, then why allow a few malcontents to put the burned-out light bulbs back in the sockets?

Getting Them Out of the Cave

Statistically, there will always be some people who offer the largest and loudest resistance to change. Some people refer to this group as Citizens Against Virtually Everything (CAVE) people. Give them voice—certainly allow them to vocalize their concern—but don't allow them to take control through active or passive resistance.

Figure 10-1 shows a normal distribution of people in the workplace. The largest percentage are watchers who tend to wait and see which way the wind blows before they choose a direction to follow. What they are watching and listening for is management's true behaviors, intentions, vision, and values to be demonstrated, not just what management says it wants to have happen.

If the watchers observe the early adopters getting all the attention, they will naturally migrate toward that behavior. If, on the

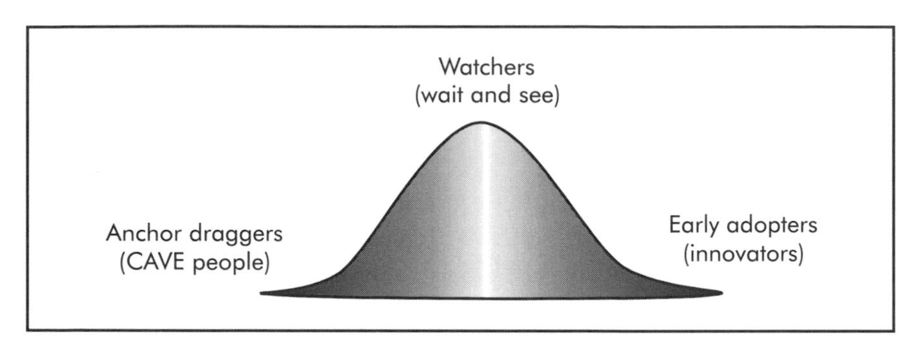

Figure 10-1. Normal distribution of attitudes in an organization.

other hand, management spends all its time with the more vocal anchor draggers, then more resistance to change will be manifested in the attitudes of the watchers.

We all want attention, and will model our behavior on what gets us that attention. Spend time with the early adopters. By doing so, the whole distribution curve can be moved toward the right side of the scale. You can not make an anchor-dragger happy anyway, so your time will be wasted in trying to salvage that sinking ship, and in the process, all the watchers will be migrating toward it.

Engaging and fully mobilizing the intellectual resources of everyone in the company is truly the variable that separates a World-Class from an ordinary company. Every company has much the same resources. The only resource that truly sets one apart from the rest is its people. People are the greatest variable in the mix. All other things being equal, the policies, leadership (coaching), and individual dedication and willingness to participate (team dynamics) are ultimately what sets one company (or team) apart from the field.

Why Didn't Ray Keep it All for Himself?

By now, the Ray Crock story is old news—how this milk-shake-machine salesman talked a couple of brothers (the McDonalds) into allowing him to sell franchises of their fledgling fast-food chain. The essence of this story is that Crock could have tried to run 12, 24, or 48 stores himself, but he had the insight that only so much of his time was available. If he wanted to grow the business, he was going to have to involve others, and if he wanted them to be invested in the success of the business, they had to buy into it. The idea of a franchise was certainly not new, but Crock took it to a new level. He did not simply empower people; he enfranchised them.

Yet, the franchises still had to operate within the limits of a well-defined set of guidelines. Although they had a measure of autonomy, they were expected to manage and deliver a specific product in a certain way. They were free—and strongly encouraged—to explore ideas for improvement because it was their business. At the same time, they were not free to go off willy-nilly in search of new products that the parent company did not support. Franchise owners had a certain latitude, and they knew it was in their best interest to go only so far with that freedom.

To get people to buy into the franchise idea, Crock had to define a particular set of needs that met the personal criteria of each potential entrepreneur. First, there was the need to feel valued emotionally as well as being rewarded financially for participating. Added to that recipe was the need to be treated with respect, the need for a certain level of autonomy, and the need for personal growth. If this was to happen, Crock had to be willing to share.

Guess what? If you want a team of free-thinking, high-energy, entrepreneur-minded employees, that list of needs is the same: respect; being valued; being well compensated; having a measure of autonomy; having discretion to make some important decisions; being provided a means of personally growing in knowledge and skill; and ultimately, gaining a high degree of self esteem.

The issues are the same as those Ray Crock had to address. You get what you give. If you are not willing to share, don't expect much. There must be shared involvement, and mental and emotional ownership.

SELLING IT TO THE BEAN COUNTERS

Six-Sigma and Lean-Manufacturing initiatives are not at odds with the idea of making money. However, you might think so if you have sat through a meeting where a continuous-improvement team had to try to convince the resident bean counters that a $5,000 set-up reduction project was a good use of the organizations' resources.

The short-term/long-term argument usually begins on the first day a Lean-Manufacturing engineer sets foot in the building. It seems counter-intuitive to spend money on a non-capital item. You can't sell a Design of Experiments project or Kaizen event like you would a machine tool, particularly when the payback may seem like trying to sell a "pig-in-a-poke."

So, how can a Six Sigma or Lean project be sold to the finance people, and how can the finance team play a more meaningful role in facilitating positive change?

Agreeing on the ground rules will help. What exactly will be measured? What can the team consider the financial benefit? For example, when a team saves an hour per day on a set-up reduction project, do they get to calculate the opportunity of reducing inventories? Or, being able to run the process an extra hour every day for 52 weeks per year? Alternatively, will the team be hamstrung by having to justify its recommendation strictly on labor savings? How about the value of shortened lead times or the dollar value of reduced floor space? Is the team allowed to show such benefits or is it given credit only if there is an immediate cost avoidance by using that space for another product or project right away? Having clearly defined rules on these issues will help avoid a lot of head butting.

Maybe We Should Just Rob Banks!

Thinking about it short term, it would probably be easier to rob banks than work hard at a regular job day after day. If the pros and cons of these two choices were objectively mapped out, it might actually make short-term financial sense to rob rather than work. That is why criminals do it. To the short-term thinking criminal, it must seem stupid to work hard day after day to earn money rather than just take it. If the same crude financial approach were taken to business, the initiative to get better would never be taken. All the tools of Six Sigma and Kaizen take time, effort, and resources. They require long-term thinking.

So, how can financial management be convinced to fund a team recommendation? Financial-resource managers have to think outside the box, and possess far-sighted vision to realize the long-term benefits and avoid foot dragging. Even given the best data possible, every single item on a Kaizen team's list of recommendations may not be cost justified. Sometimes a company has to do something simply because it's the "right thing to do," especially when it's not fully justified financially and totally risk free. It takes a light-bulb moment. Continuous-improvement projects can never be fully cost justified without a company's long-term thinkers shining brightly. This is not to say that questioning a proposal is not a worthwhile exercise. The team should be challenged to prove that there is a payback for its recommendation. A simple return on investment (ROI) spreadsheet can often answer many of the concerns from those responsible for releasing funds. It can also help support the team's arguments for adopting the proposal if it contains meaningful and logical elements. Table 10-1 is an example of a simple return-on-investment work sheet.

Go for the Bread and Butter

"Bread and butter" jobs are frequently repeated jobs or ones that flow through the plant in the same sequence, with roughly

Table 10-1. Simple Return-on-Investment (ROI) work sheet

Item	Benefit	Cost	ROI
New roller table	10% reduction in run time, saves 39 min/day = 169 hr/yr, $2,535 labor cost/year	$1,500	0.59/year (7 months)
Reconfigure cell, remove racks	3,000 ft^2 (279 m^2) now available, cost avoidance, do not need to add space. Industrial floor space worth $15/ft^2 (1.4 m^2) equal to $45,000 value.	$13,750	0.3/yr (3.5 months)
Total	$47,535	$15,250	0.32/yr (3.8 months)

the same labor and machine requirements. Nice work if you can get it, right?

At the other end of the spectrum is the customer who brings in the odd item or job on an infrequent basis. He always needs a modification or two, never has enough information, and the specifications are less than clear. Staff must spend an inordinate amount of time chasing down missing numbers or fuzzy details. These types of jobs are called "dogs and cats" because they are unpredictable, and they usually fight for priority as they move through the shop like a pack of strays. There are times when customers like these hold the small shop back from growing business with customers of the bread and butter variety.

One job shop got to the point (about $20 million in sales) where the $50,000-or-less-per-year customers were creating more havoc than revenue. Therefore, it developed a plan to match some of its customers with the job shop's smaller competitors. The job shop smoothly transferred each client to the new vendor by providing prints, CNC programs, etc. In most cases, all three parties are now happier. This critical step required serious thought. It didn't seem to make sense that the company was giving away work in the face of a weakening economy. However, dealing with so many dogs and cats was creating engineering and programming constraints. The job shop now had time to concentrate on its core customers.

Another light-bulb idea is for salespeople to start unlearning the bad habit of selling like crazy at the end of the month, quarter, or fiscal year. This creates tidal waves of work that wash over everybody in engineering, order entry, production planning, and manufacturing. Before long, customers will have been trained to economize by buying in large batches. They need to know that due to shorter set-up times, faster lead times, and lower inventories, there is no longer a financial incentive to wait for that "end-of-the-month special."

IT'S ALL ABOUT FLOW

The concept of one-piece-flow makes perfect sense with an assembly line making one car at a time. However, in a job shop, the vision of perfection looks more like flow rather than one-piece-flow. Being able to stand back and see all material moving rather than languishing around in racks or sitting in warehouses is the goal.

It should take no more than 10 times the value-added time to process an order. For example, if it requires 10 hours of labor to produce a work order, then the total lead-time for processing an order through the manufacturing environment should be no more than 100 hours. This is certainly an aggressive goal, and it takes most companies some time to reach the goal of a 1:10 value-added ratio. But the longer a company waits to get started, the harder the transformation. If a company waits six months, it will be simply six months behind where it could have been (and possibly six months behind the competition).

Measure the Right Things

How can value-stream managers be measured for effectiveness? Machine run time? Labor allocation? Labor cost per unit? What is the right measurement? Instead of "How are managers measured?" maybe a better question is "What is the goal of the company's manufacturing managers?" An even better question might be: "What is the goal of the manufacturing process and what should be happening?"

Regardless of the product or service, the answer to that last question should be: "Get it on the truck at the lowest possible price, at the quality and on the delivery schedule that satisfies the customer, and get paid a reasonable price for doing so." Manufacturing managers should be focused on getting it on the truck, instead of spending their time performing non-value added activities like filling out work-center uptime reports and machine-optimization studies.

Using a sheet-metal example, an average turret punching machine might have 50 tool stations. The company has paid for each station and for each tool installed in each station. Yet, only one station can be used at a time. Should the company keep track of the efficiency of each station? If it did, the efficiency might be something like 2%. Should the company try to figure out how to capitalize and optimize the remaining 49 stations that are not being used? Of course not. So why spend time trying to figure out how to keep machines running just because they occupy space in the shop. Shouldn't the question of running a machine or not be based on whether or not it will help the rest of the manufacturing process get product on the truck?

The manager of a racecar team does not have the team run the car at full speed the night before the race just because there is fuel in the tank. In manufacturing, this is called *over-production*. It makes no economic sense to produce something that nobody needs or wants just to keep a machine running.

Some alternative metrics to use for measuring managers' effectiveness might be:

- on-time delivery performance;
- value-added ratio (time spent working on materials versus lead time);
- set-up times compared to targets;
- number of operations having standard work defined;
- work-order lead times compared to target;
- employee retention;
- team morale;
- productivity (sales dollars or cost of goods sold) per team member;
- team attendance;
- team cross-functionality (depth of knowledge);
- yield; and
- quality (defects per $1,000 sales).

Plan for Production

How valuable is time spent using job-scheduling, job-tracking, and job-costing software? What does a company get for the money it invests? Who really needs the information and why? What alternatives are there to these tools?

How much does it cost to generate a work order? If there are three production planners, each making $33,333/year, then the annual cost to plan jobs for the shop is $100,000. If this team generates 400 work orders per month (or 4,800 per year), then the cost per work order is $20.83. A recent study has shown this number averages closer to $35 nationwide.

Why do people need a work order? It tells them what to make, when to make it, how much to make, and to some degree, how to make it. The real reason that so many material-management systems are sold is that over time, companies develop longer and longer lead times. This requires that someone in the office know where the material is at any given moment in time (just in case the customer calls). The computer system ends up being the tail wagging the dog, because now, instead of operating his machine, the machinist is standing in line to log in or out of a job. Instead of material flowing seamlessly and rapidly through the plant, it waits in queue because the system needs to locate it, and a queue between operations is the only way to identify its progress.

What are the alternatives? If there are to be meaningful choices, the options must still answer the questions: what to make, when to make it, how many to make, and how to make it. Then the big question is: "Did the company make any money producing and selling this product?"

Visual signals or displays for moving or making a product can work well in highly repetitive operations like those at Toyota. Two-bin systems that virtually eliminate the need for paper work orders seem to function well in plants like Harley Davidson. The bins themselves manage in-process inventory. As the operator empties one bin, he or she begins working out of a second bin. In the meantime, the empty bin is sent to a machine shop or fabrication shop to be refilled. All the information about when it is needed, how many, and specifications are included within the bin itself or on an accompanying card.

Track the Non-repeat Order

How can a job shop hope to reduce the need for work orders and tracking mechanisms when the product is not a repeat order?

The reason that most companies finally decide to buy a material requirement or procurement system is that they are set up to operate in functional departments because of growth. They group all customer jobs into a big hopper and release the orders to the shop in the best order that will optimize equipment and labor. Then, they stand back and hope the jobs come out the bottom of the funnel in the correct sequence. When it doesn't work, they try to police the process by adding another layer of cost and lead time using job-tracking software packages. This theoretically takes the decision-making process away from people (who are always trying to manipulate it to make themselves look good at the end of the week, month, or quarter), and gives it to an unfeeling, unemotional computer. Is this a cure-all? What does the company really get for its money? Often, companies get:

- everyone standing in line to log-in and log-out of jobs;
- to buy, copy, and store reams of paperwork;
- to enter millions of keystrokes per year;
- to add maintenance, hardware, and software costs to overhead;
- to spend thousands of labor hours training and retraining; and
- to have questionable or unreliable reports and unbelievable costing analyses.

It is not fair to offer criticism without suggesting an alternative or two. In regard to finding out the true cost of manufacturing a product: if you want to know how much something costs to make, go audit it. A full-time auditor costs a lot less than seven people spending half their time entering pencil-whipped data into a costing system.

If the job was estimated to take one hour and the operator actually completes it in 30 minutes, there's a good chance the operator will find a way to apply the extra 30 minutes to the next job. So how good is all the data?

By dividing the total pile of work done each month into manageable buckets, value streams can be developed to deal with certain kinds of products. By segregation, teams can better plan and produce, sometimes by using recycled work orders, limiting finished-goods inventory, or directly communicating with the customer's assembly line.

There should be no misunderstanding. A material requirement or procurement system can serve a purpose for solving some macro planning issues like "When will another machine be needed?" or "Will the company need to hire additional people next month?" On the other hand, micro planning often can be simplified by dividing the order file into manageable buckets of work for the value streams. Cellular layouts often can be designed to look, feel, and operate as a single machine. This greatly simplifies the tracking process, reducing the need for complex computer systems.

Strive for Manufacturability

Re-inventing the engineering and manufacturing information flow for operations is like gathering low-hanging fruit, yet it is often overlooked in favor of things that seem more visual. The trouble is, what gets fixed are often cosmetic rather than systemic problems.

Thoughtful redesign of CNC programming practices and standardization (that is, programming to a standard turret) are examples of improvement opportunities. These can yield cost savings and improve quality not just once, but every time the program is run on the machine, especially with smaller lot sizes.

Anticipating some essential variables, such as excessive set-up time, rejects, and wasted motion, which manifest themselves later as quality problems, can yield unbelievable financial rewards. Set-up reduction is not just the responsibility of the machine operator. Having engineers and programmers trained to think in terms of fast set-up (or no set-up) and manufacturability is where the real responsibility belongs, and where the real gains are to be found.

Challenge the Machine Operator

Machine operators must take up the continuous improvement challenge. There is an enormous need for skilled and knowledgeable machine operators who are World-Class disciples.

Companies that are serious about implementing Lean initiatives often allocate 1% or more of the sales to sustaining the Kaizen promotion office, continuous-improvement teams, and total-productive-maintenance programs.

THE PROFIT IS IN THE PITS

For years, it made perfect financial sense for a company to run the largest batch size possible and put that material in stock to sell and ship as customers needed it. If the customer wanted 1,000 units per month, there was no value seen in setting the machine up four times to run 250 per week instead of all at one time. To run off only 50 per day per set-up seemed total lunacy.

Assume a shop sells $10,000 of product per week and the material cost is 40% of the selling price. The cost pie chart might look something like the one shown in Figure 10-2.

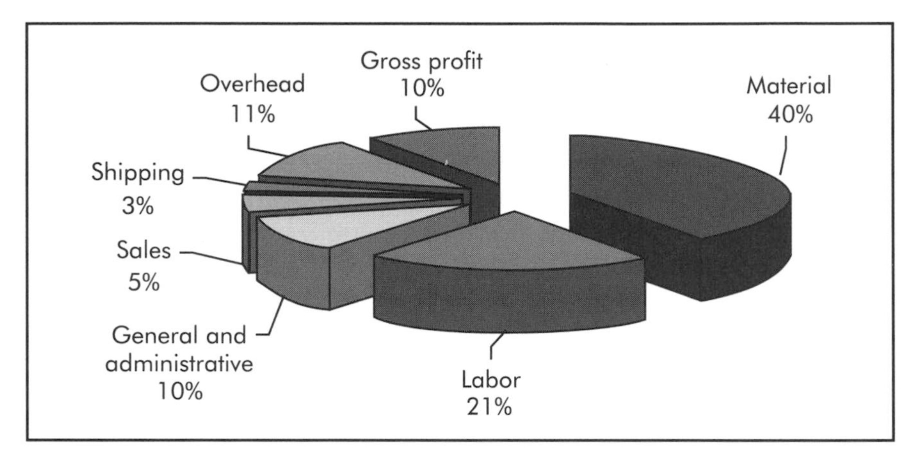

Figure 10-2. Cost/profit chart.

In any business, there are certain constants regardless of the product made. The more times inventory is turned over each day, week, or month, the better the return on investment. Taking costs out of the cost pie can have the same effect on the bottom line as increased sales with no additional capital investment. In the cost-pie example, assume a 20% increase in labor productivity is found along with a 10% reduction in overhead costs (inventory, machines, power, consumables, etc.), as shown in Figure 10-3.

The gross profit effect of the investment (profit at 15% instead of 10%) is now the same as if $15,000 of product had been sold instead of $10,000. Actually, it is even better than that because if the company really could sell $5,000 more per week, it would have to hire more people and possibly buy more equipment. Cost avoidance is one of the immeasurable benefits that Dr. Deming talked about when trying to convince American managers to think creatively and holistically when adopting improvement initiatives.

One of the greatest benefits that an operator can offer his or her company is to help management rethink and re-engineer the

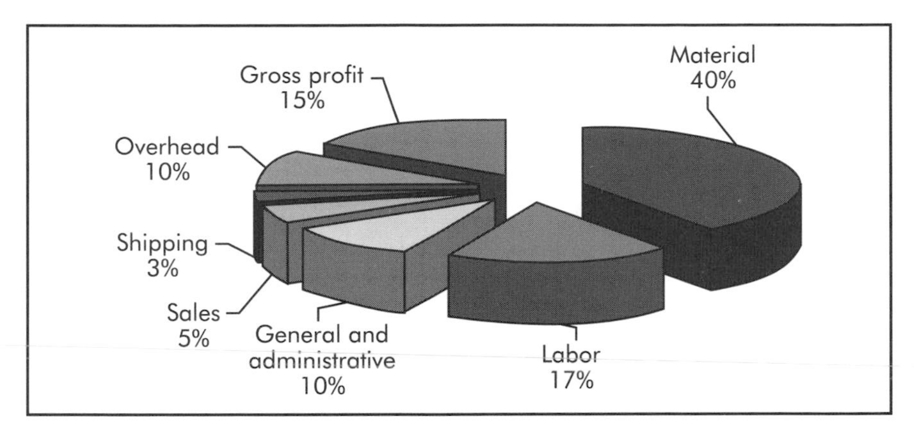

Figure 10-3. Improved cost/profit chart.

process used to generate product. If an operator were to examine what is done each day, he or she would generally find a large amount of time spent doing non-value-added activities. For example, there is time spent performing activities like logging in and out of jobs, moving material into and away from a workstation, setting up machines, or trying to find the supervisor or a quality-assurance person to perform a buy-off or first-article inspection on a new set-up. All these activities may be necessary, but they are non-value-added. What is meant by non-value-added? *Value-added activities* are those activities a customer is willing to pay the company to do: like painting, welding, bending, or punching. All these activities add value. Moving, counting, ticketing, etc., do not physically change the part, and therefore, add no value from the customers' perspective.

What about set-up? Surely, the customer wants the company to set up the machine correctly. This is where the light-bulb moment needs to happen for the operator. For a racecar team, the pit stop is not where the money is made, but where it is lost. The goal is to get the car out of the pit safely and as soon as possible. Fortunes are won and lost in the pits. So how come a company's machines sit in set-up mode for 10, 20, 30, 40 minutes or longer while tools are changed, material is moved in and out, operators search for paperwork, etc.?

Set-up reduction can help drive the profitability of the entire organization. Some of the activities that happen at work cannot easily be improved or avoided. However, set-ups always can be reduced. There may be some costs and time involved, but most set-ups can be reduced by at least half. This can be done by simple reorganization and reassignment of some of the more mundane tasks to someone else or by preparing the set-up off-line while the machine is still running.

A Kaizen event should be run to reduce set-up time on the bottleneck machines first. The minimum goal for job shops should be to cut that time in half. Next, the company should aim for single-digit set-up times (set-up times that average less than 10 minutes).

Why not Trade Expertise?

Although it is nice to have total control, sometimes using the expertise of specialists outside an organization can have beneficial and hidden advantages. Certain things are simply less expensive or more reasonable to have an expert do. A company that tries to be all things to all people—can actually hold it back from capitalizing on its core competencies.

However, depending on someone outside an organization always adds a level of complexity. One solution is helping educate vendors in the principles of Six Sigma and Lean Manufacturing. Helping them improve their process will ultimately help the small company. Bringing supplier representatives into Kaizen events, workshops, or in-house training sessions is a good way for them to learn about what the company is trying to do. (What can they do? Turn down a good customer's invitation?) There should not be any problem convincing them that participating in these events is in their best interest. Seeing the company's commitment to the process will help the light bulb come on for them as well.

How can the company select the vendors who are most likely to support its World-Class initiatives? Just like hiring people off the street, it probably won't find the ideal candidate. Coaching is generally needed, as well as a good deal of patience and relationship building. If an employee or a supplier senses that a company has their best interests at heart and that it wants to see them succeed, they will generally perform at a level higher than might normally be expected.

Taking Inventory—No, You Take It!

What about inventory management? There are countless ways to gain knowledge of inventory-management techniques, including American Production and Inventory Control Society (APICS) training and other like training programs. Most software companies include (or sell) training packages related to their resource-planning modules.

Inventory is not the disease. The disease is excessive lead time. Inventory is simply a mechanism that companies use to cover up or hide the symptoms and the disease. If a company has a four-week lead time, then logically there is at least four-weeks worth of inventory sitting around somewhere. How can lead-times be cut in half? Cut every set-up in half. Cut every lot size in half. Then, inventory will be cut, resulting in lead time cuts.

One of the light-bulb moments for management is to stop thinking about inventory strictly as a dollar figure, and begin thinking about it as time. If inventory can be cut in half, lead time can be cut in half.

Getting vendors to begin managing their own inventory is another tool forward-thinking companies use to save administrative costs and time. For example, buy-out items like nuts, bolts, screws, weather-stripping, etc., can often be supplied and labeled in such a manner that the vendor still owns it until the team pulls off the label and faxes it to the vendor. No transaction is necessary until the company actually consumes the material.

A new trend in procurement networking is having the supplier become a more permanent fixture in the supply chain. The buyer supplies forecast information and frequently shares ideas that can lead to mutually beneficial financial improvements. Such results have encouraged many companies to adopt a more meaningful and secure relationship with their suppliers—clearly a win-win relationship.

Quality's Identity Crisis

The title of most company's quality-function leadership position has changed many times over the last few years. In the 1970s, quality policing activities were performed by a quality control manager; the 1980s saw the title on business cards change to "quality assurance manager." This title was more in keeping with the prevention versus detection role many companies had adopted. "Quality system manager" titles were doled out during the 1990s and better fit the mold of the ISO 9000 structures that began sweeping the globe during that decade. Finally, the new millennium has seen the more descriptive and all-encompassing title "director of quality." This is a title for someone charged with:

- maintaining compliance to third-party assessment,
- developing and administrating in-house education,
- conducting internal audits,
- coordinating Six-Sigma team activities,
- developing cost-of-quality parameters, and
- training vendors in the preventive and corrective-action disciplines used by World-Class organizations.

Lean-Manufacturing techniques like 5-S (organizational housekeeping), best practices, standard work, and others lend themselves well to enhancing the operator's ability to build quality into the process. Many directors of quality are just now discovering the synergistic power that Six Sigma, Lean Manufacturing, and Total Quality Management can have for the organization. Time and experience have shown that overall quality costs can be reduced, and the level of defects can, at a minimum, be cut in half every year when applying these tools in a holistic fashion.

Concurrent Developments

There are mixed opinions about research and development. Some companies can be so R&D driven that they bleed to death by chasing any and all new products and technologies, while never spending enough time or resources to become better at what they currently sell. Yet, there is clearly a need to look to the future. There may not be an answer to what is the right amount of time and resources to apply to the development of new markets or how many "pie-in-the-sky" ideas to chase. There must be a balance between what is ideal and what is real.

The idea of developing and introducing a new product with full participation from marketing, engineering, manufacturing, production, and packaging, is not new. The goal of concurrent or simultaneous engineering is to have high quality and fully engineered drawings and specifications well thought out beforehand, along with well-defined manufacturing processes, and even suggested inspection, testing, or simulation processes. All of this occurs before shop drawings are released to the manufacturing floor.

A Value Stream Mapping exercise is just as important here as it is on the manufacturing floor. To make the process and goal visual for people, there are a couple of techniques that can be applied. One is shown in Figure 10-4. It is an example of how a current new-product introduction process can be reduced in the future.

The term "concurrent engineering" is probably not new to you. To more fully understand the details and steps necessary to be

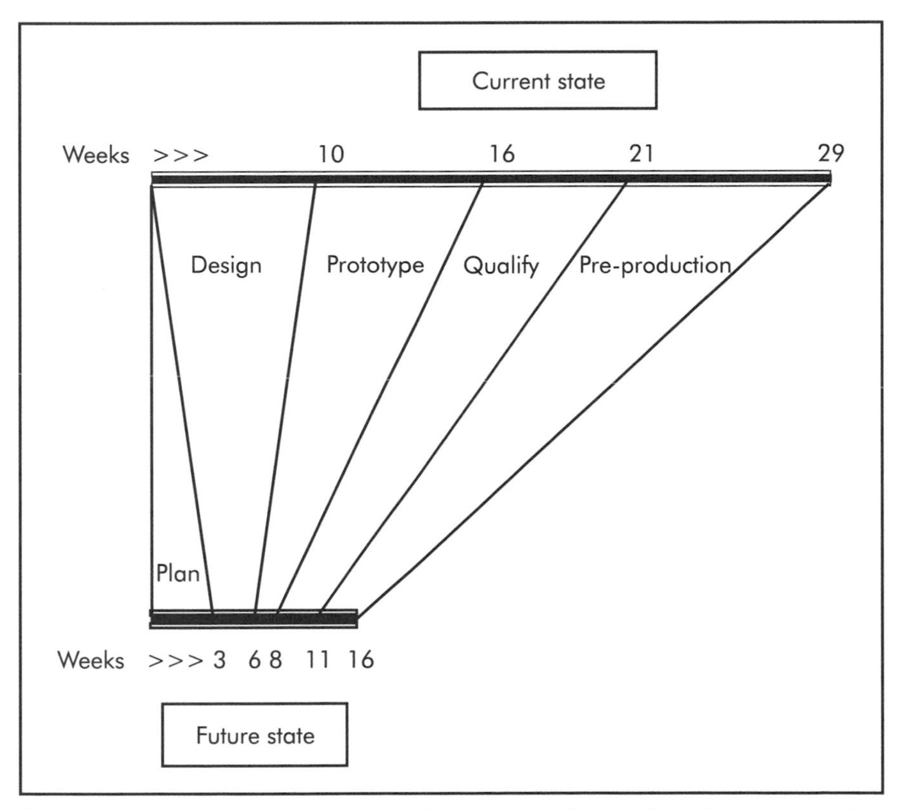

Figure 10-4. Current state versus future state for engineering-process improvements.

able to perform this technique well, see Appendix B, a sample concurrent-engineering process. This process was mapped and developed to streamline and ensure that each critical step is managed from marketing concept to delivery. This is a generic example with elements that probably do not apply to every organization. While the example in Appendix B more closely models an original equipment manufacturer (OEM) format of design from marketing through pre-production, the steps for a supplier are also presented. They could be used to show how the customer and supplier should interact in a true concurrent-engineering program.

Hotshot Teams for the Hot Jobs

Some companies are setting themselves apart by offering 24-hour turnarounds on prototypes and emergency part manufacturing. This kind of offering is a powerful way to gain loyalty from customers who do not have their own fabrication capability or capacity to rush materials through. These companies have set up special teams of "hotshots," who operate within the normal structure until a hot job comes in, and then they spring into action, taking the project from engineering through assembly and delivery, if necessary.

While skirting many of the normal processes, the team develops traceable and clear documentation so that if the job repeats or is revised and run again, it does not have to completely reinvent everything. The outputs may be hand-drawn rather than computer-aided-design (CAD) rendered, and the work order may be generated in the form of a checklist rather than from a computer-driven system.

There are many benefits besides quicker turnarounds. The hot job is usually very disruptive. Using a cross-functional team helps keep questions and disruptions to the rest of the shop to a minimum. Normal bread-and-butter jobs continue flowing.

RECRUIT, RESOURCE, TRAIN, AND RETAIN

Although the issues of recruiting, training, and retaining are not the sole responsibility of the human resources (HR) department, it is their job to ensure that managers and employees have the right tools to execute a sound HR policy. What follows may be some of the most important information in this text, because it is fundamental to the core vision and values embedded in any truly World-Class organization. For example, just because a company employs 100 people with an average tenure of 10 years of service doesn't mean that it is reaping the rewards of 1,000 years of their combined wisdom and experience. People may have knowledge, but just like an orchard full of peaches, the fruit has to be harvested if others are to benefit. How can this combined knowledge be harvested, turned into wisdom, and used to advantage?

The foremost role of the HR manager is to help others fully recognize that the human element holds the greatest potential for distancing any organization from the competition. Full mobilization of the intellectual resources is necessary.

Practice Knowledge Control

One company developed an in-house college to train and share knowledge so that employees would not have to drive 60 miles (97 km) round trip for evening classes such as blueprint reading. Finding someone to teach the math class was a problem until it was found that one of the machine operators was just a class or two away from a masters degree in mathematics. Many employees were willing to act as in-house instructors, so a train-the-trainer program was conducted. They received compensation for the time they spent teaching, and the company ended up with over 30 in-house programs, from blueprint reading to laser operation.

All you really have to do is ask! You may have to be a bit of a detective or gold miner to recognize hidden knowledge and remove the reasons that typically cause people to hide their knowledge rather than sharing it with others.

Many companies drift along rather unencumbered by process problems until an essential operator leaves or gets sick, and then they suddenly realize that this person was the only one with the knowledge to do a particular task. "Tribal knowledge" is a situation when information is conveyed by word of mouth, or held secret by the leaders of the group. When leaders depart or are absent, the remaining tribe members struggle or are doomed to failure because the depth of knowledge is so poor.

The 2000 version of the ISO 9000 criteria requires companies to give a higher level of attention to knowledge control than in the past. Just like process control, the control of knowledge ensures that customer satisfaction is not at risk just because someone goes on vacation or maternity leave. Again, documentation and training is required.

The light-bulb moment, in this case, resulted from a process founded on trust. People (including you) are not willing to share knowledge if they are made to feel de-valued because they have shared their knowledge and now more people know what only they used to know.

HR managers are paid to help the company get better every day, yet sometimes they say: "What if we spend all this time training them and then they leave?" The best response to that question may be another: "What if we don't train them and they stay?"

Get in the Training Business

For the company (not just individuals) to grow and prosper, the best source of training exists for the most part within the walls of the organization. The company possesses the tools, materials, machines, parts, assemblies, and processes that no college in the world can duplicate. It also has the most important training mechanism: the people who know the processes and customers.

Yet, company management may think: "We are not in the training business. We are manufacturers. How can we hope to develop trainers out of operators?"

First, a couple of key questions have to be answered. How do you get people to be willing to share knowledge that may have taken them a lifetime to acquire? What values drive the willingness to share?

People who feel secure and valued are more willing to share than those who feel threatened that the moment someone else can do their job they will be de-valued. Company management must create an environment where people feel proud to be a mentor and are rewarded (and in some cases paid) for doing so.

There also has to be a process in place. For example, the government requires that pilots use a preflight checklist before the plane departs. Even if the pilot has flown the same plane everyday for his or her entire career, that checklist must be used every time. The same is true of training. If accidents, mistakes, and injuries are to be avoided, there must be a documented process. It doesn't have to be formal, but it has to have enough structure so that no critical training element is left to chance or memory, or left out because Bo thought Moe already went over that material with Joe.

Mine that Gold

There is the choice of doing things the old complicated way or finding an easier way. Making things more difficult will guarantee that people will scrap any new idea at their first opportunity; it soon will be a faint memory. When developing new policies and procedures, keep in mind that change comes hard enough without adding the self-destruct mechanism of complexity.

Knowledge is golden, so go to the office supply store and buy 25 of those gold-colored cardboard banker boxes (the kind that important papers are stored in). Assign someone who enjoys talking to people to interview the best operators, using the checklist in Table 10-2 as an interviewer might. As the interview is conducted, leading questions are asked about what new operators need to know on their first day, first week, first month, etc. An example of a partially filled in form can be found in Appendix C. This outline is then crafted into a curriculum complete with samples, photographs, and associated paperwork and methods of testing for knowledge. This process can usually take place in a couple of 45-minute sessions.

Once filled out, organize the items on the list into a logical training sequence. Then, have department personnel find samples, examples, blueprints, broken tooling, models, part profiles, old and new tooling photographs, videos and sketches of each item needing to be trained, and throw them all into a banker box. See Appendix C for an example of a basic training outline for several skill levels (for example, first week, second week).

Now, anytime a new employee needs to be trained, an expert can go to the shelf, pull down the banker box, reach in, retrieve the training checklist, and then train the person in every critical step of the operation.

Table 10-2. In-house training curriculum checklist (interview form)

- ❒ Department
- ❒ Machine ID
- ❒ Interviewee
- ❒ Interviewer
- ❒ Variations of this machine
- ❒ Hand tools
- ❒ Measurement tools
- ❒ Math requirements
- ❒ Safety concerns
- ❒ Specifications/documentation
- ❒ Materials
- ❒ Tooling (location, storage, handling, loading/unloading)
- ❒ Quality requirements and concerns
- ❒ Equipment nomenclature
- ❒ Machine adjustments
- ❒ Basic skills (first day)
- ❒ Machine start-up and shut down
- ❒ Standard work
- ❒ Secondary skills (first week)
- ❒ Advanced skills (first month)
- ❒ Basic set-up skills
- ❒ Advanced set-up skills
- ❒ Horror stories

Flex Those Hours

An alternative to the usual one- or two-shift operation of a manufacturing company is to consider a more flexible operating schedule. Figure 10-5 shows how production capability over the course of a week looks in most shops.

What could a company do with the remainder of those 180 hours? Could it put the most repeatable jobs on the off shifts where there is less need for full-time supervision? Could it pay well

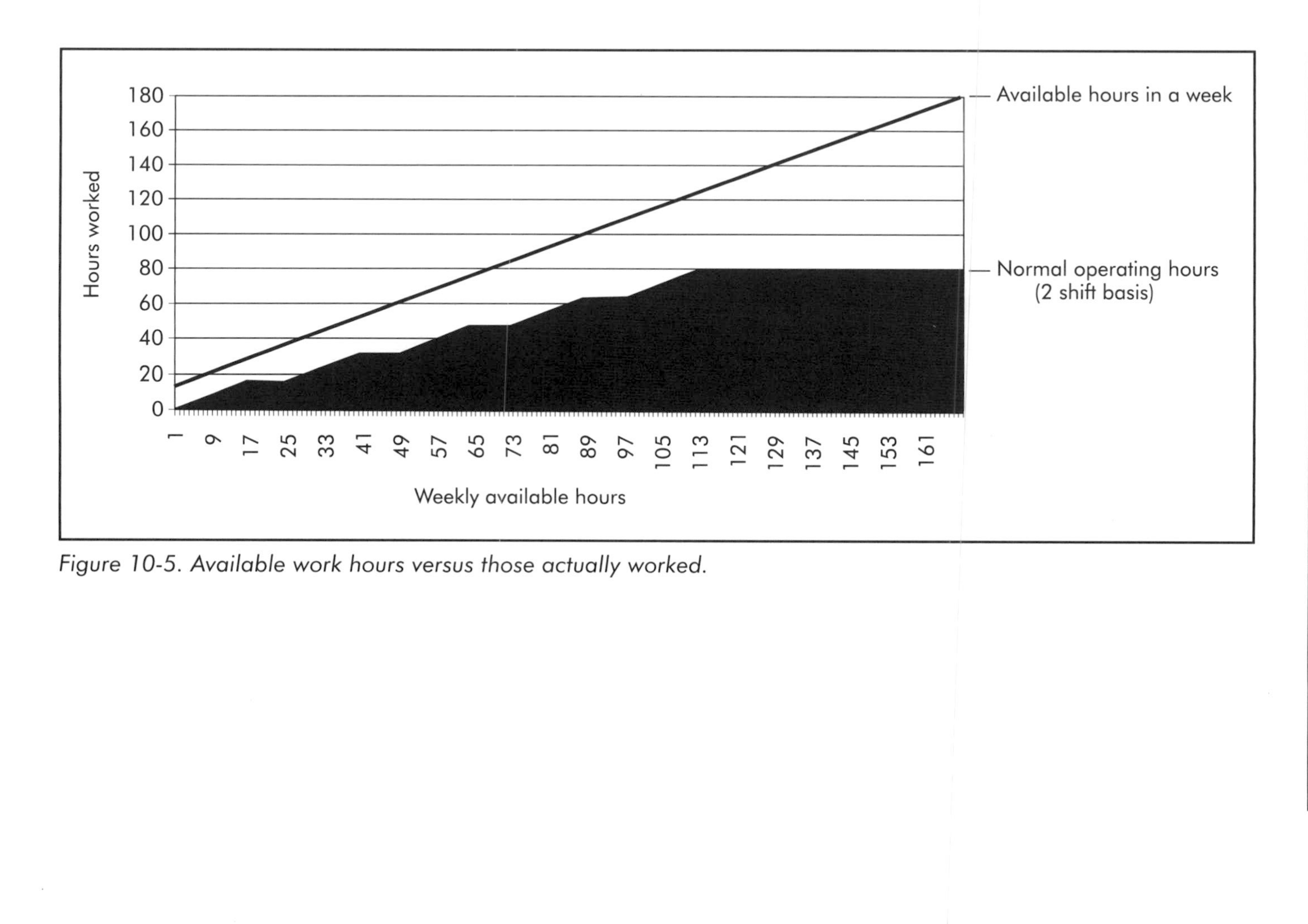

Figure 10-5. Available work hours versus those actually worked.

enough to attract the best people to cover those shifts, and gain virtually 24-hour availability? This would free up the primary shift to deal with the dogs-and-cats (prototypes and oddball materials) jobs. The people who need to be available (customer-service personnel, engineers) would be there to answer the questions that come up. It also allows unparalleled flexibility and cost avoidance. The company would gain over 100% machine capacity without buying another piece of equipment or larger buildings.

Having 100 people all working one shift equals 4,000 labor hours per week. Having 25 people working four shifts (Monday–Thursday 10-hour shifts and Friday–Sunday 12-hour shifts) allows time for maintenance as well as round-the-clock availability to meet customer needs.

If there are worries about finding people to staff these shifts, just experiment with it for a month. After people get a taste for three or four days off every week (and having the golf course or hiking trails all to themselves) you'll have a fight on your hands trying to get them to go back to a more traditional work schedule.

Anticipate Departures

What does it cost to lose a person? Estimates range from "One third the annual salary of the position being vacated" to "It depends on the tenure of the person who is leaving" to "More than anyone could imagine." There will also be the kind of departure where everyone breathes a sigh a relief. Yet, how much does it cost to lose a talented and valued team member?

In the example in Figure 10-6, it is assumed there are 450 employees, and a 33% turnover rate. It could be argued that the pieces of the pie are too small or too large, but the point is, are true costs known? In this example, to lose 150 people per year costs nearly $300,000, and this does not account for all the immeasurable costs.

Maintain Prevention

The typical job of any maintenance manager and team member is to provide machine capability, machine reliability, machine repeatability, and low levels of machine-induced variation. Another responsibility less often undertaken is to ensure that the machine

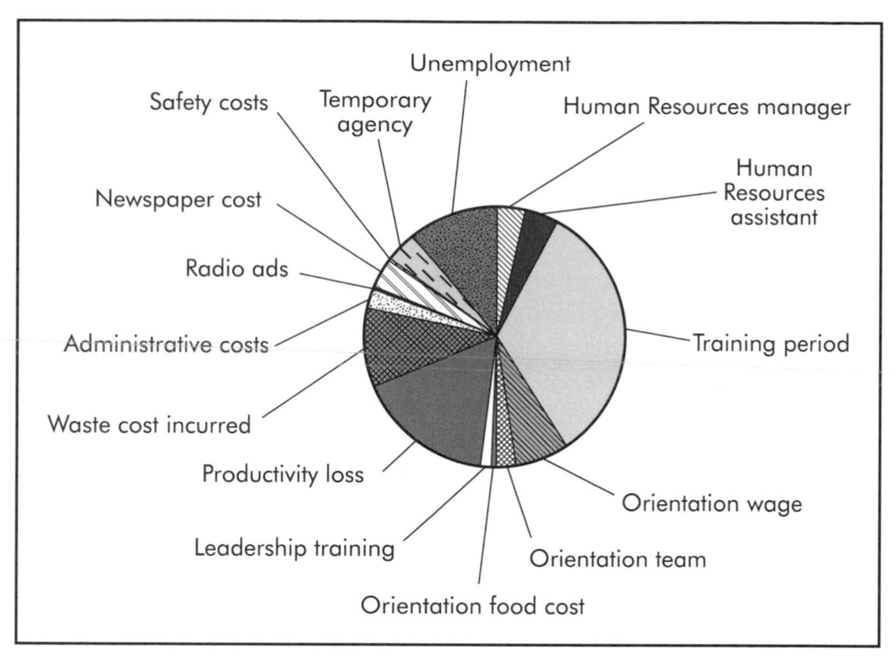

Figure 10-6. New-hire costs.

operator feels competent in assisting in the identification of wear-and-tear signals, visual inspection of key wear points, and has complete knowledge of daily machine preventive-maintenance procedures. This can be likened to the wisdom of a doctor who engages his or her patient in their own health-care decisions rather than just running in and out of the examination room. How arrogant it is when the doctor rejects input from the patient or shows little regard for the patient's opinion in the prevention, diagnosis, and treatment of his or her own body.

Where is the majority of the money to be saved in maintenance? In the prevention process. Training people to grease and maintain their machines will save downtime and repair costs.

Manufacture Flexibility

When job-shop managers travel to machine tool shows, they often see and hear the "bigger-is-better" sales pitches as the salespeople stand beside the biggest and best machines. One thing the major tool manufacturers need to hear from job shops loud and clear is that a $1 million piece of equipment running full time at maximum capacity is less attractive in the new "flexible" environment than 10 pieces of less complex equipment costing only $100,000 each. The ultimate goal is to get away from shared resources that force batch manufacturing. Machines must be easy to set-up or dedicated to work within a defined value stream.

When building a new facility or designing a new layout for an existing plant, flexibility needs to be the watchword. Centralized power bus bars to allow quick disconnects and flexible tethering to machines will allow and encourage the reconfiguration of cells when required by changes in market demand. All utilities (air, water, compressed gases, etc.) also need to be installed in a manner that facilitates relocating equipment over a weekend.

Because of long entrenched training, most manufacturing engineers tend to be very analytic by nature. They often try to dictate to the shop teams how much more effective and easy to maintain a layout is when all machines are at 90° angles to each other, and how they need to allow space for forklift traffic, and so on. Educating industrial and manufacturing engineers about the principles of Lean Manufacturing will go a long way to help them lose the perception of a need for total flexibility that is taught in some industrial-engineering programs.

Establishing a poor layout around the chance that you will have to work on a machine "sometime" (0.1% of the time) forces you to live with a poor layout 99.9% of the time. If a company tries try to meet the needs of every dog-and-cat order that the sales team drags through the door, everything is forced to go that way. Even the bread-and-butter jobs are required to flow the way of the dog-and-cat. There is never a chance to be really lean at anything.

The industrial- and manufacturing-engineering functions should be among the most well-trained Lean advocates. They should be encouraged (or required) to complete a Six-Sigma or Lean-certification course before being allowed to assist a team in a new cell layout, machine purchase, or reconfiguration. Please note: new laws regulating earthquake-proofing of machines during reconfigurations have added a new level of complexity to designing highly mobile and flexible plant layouts. Check your local and state building codes for additional information.

Deliver the Goods

If a company is more than a few miles away from its customers, then it is probably struggling with higher fuel costs, a greater demand for smaller lot sizes, and having to absorb the costs and stresses of all of these challenges.

The new "world order" for delivery will no doubt include some combination of satellite operations, pooling truckloads with competitors, new relationships with shipping companies, and many yet-to-be-discovered solutions.

In Japan, large numbers of little delivery trucks move small lots of material from supplier to customer many times a day. Such Just-In-Time delivery services in a metro area might be a great business in the new Lean economy. Duplicating what worked in Japan seems to makes sense where vendor and customer are less than a few hours apart.

SOME MAJOR CONCLUSIONS

The high demand for energy from all the new light bulbs may require a company to be figuratively rewired.

Once aware of something that went unnoticed before, there is an unconscious mental alarm that goes off whenever it is seen. For example, assume a friend is interested in buying a new mountain bike. For months, you may have been driving past the local bicycle store and never paid it much attention. Yet, once it is of new importance to you or someone close to you, you will notice every bicycle on the road. You have a new synapse, a new biological and electrical connection in your brain. You've been rewired to notice bicycles.

Everyone's brain needs to be rewired to see waste in the company. Everyone must see it, hear it, and smell it. No one should ever step in it, walk by it, or sweep it under the rug. It must be picked up and put it on display so it is visible to everyone else and can be avoided in the future.

Also, every resource—materials, power, utilities, consumables, people, money, inventory, machines, and time—has to be metered out so that managing one (seeking optimization) does not create unacceptable waste of another (suboptimization).

There is a need for new and creative thinking within the company. This book does not presume to provide all the answers. However, it has asked some key questions that can help generate creative thought within a team. In turn, it can help move an organization to find the answers that will work in your unique applications.

Consistent use of the Define, Measure, Analyze, Improve, and Control (DMAIC) cycle process will help drive continuity of effort and create a common understanding of this time-tested process for implementing positive change:

- Teach teams to clearly define the issues and communicate the opportunities. This will put them in a better position to prioritize where their efforts should be focused.
- Educate everyone to measure the variables in their processes.
- Foster an environment where everyone is responsible to analyze the cause and effects of variables as part of their daily activities.
- Teach everyone that the vision to be a Six-Sigma performer relies on 100% participation and willingness to improve daily and reduce variation in every activity from conceptual design to shipping. Use the tools of brainstorming and invite creative thinking. Remove the barriers of old-world thinkers and managers who refuse to capitalize on the idea generation techniques of the group. New construction often requires demolishing old buildings, clearing away the "concrete heads," or else nothing new can happen.
- Control and sustain the progress through discipline, but not punitively. To modify and control a process there must be clear understanding of why the change is important. Otherwise, the gains will easily evaporate. The root word in discipline is disciple. A disciple is a learner. To be a disciple assumes that there is a teacher. The job of managers today has changed. No longer are they simply directors barking out orders, but they must be teachers. That includes being a teacher of the entire DMAIC process.

Create and then Simulate

During nearly every manufacturing-cell reconfiguration project, simulation of some kind is run to redistribute the work content and balance the line. After setting up what are thought to be adequate standards for work levels, after all the calculations, there are still a few unknowns that only running the system will show.

In the past, every form of simulation has been used: shuffling pieces of paper representing parts, passing paper clips across lunchroom tables, and trading handfuls of screws, washers, and nuts to simulate a mixed-model line. Masking tape and building blocks have been used to simulate Kanban locations and point-of-use supermarkets. Mockup machines have been built from 2-by-2s and duct tape to visualize how people will move around and share work centers. Yet, none of these tools can help answer all the questions. Eventually, the hand has to be put onto the switch to turn it on. Simulations can help prepare a team for what to expect. Simulation activity is as important as a training tool as it is for testing ideas.

There are some software packages that allow testing ideas on the computer before moving equipment around. One of the more intuitive simulation tools is by Pro-Model®.

Sometimes We Walk Like a Frog

A transformation like Six Sigma or Lean Manufacturing (or any other improvement initiative) comes in fits and jerks . . . like a frog walking. Of course, frogs don't really walk; they hop. Once a frog was trying to climb a waterfall in a backyard pond. He would jump from rock to rock, and half the time he was washed back by the oncoming water, but he kept at it until he reached the top.

Just remember that two steps or hops forward and one back is better than no forward progress at all. Just keep hopping! Don't let minor setbacks be discouraging. It is never going to be easy.

World-class athletes cannot afford to stop honing their skills if they hope to stay in competition. Professional athletes have relatively short careers compared to the duration of most manufacturing careers, so they must think short term. Thankfully, in manufacturing there is the luxury of spending more time in the game.

Look at the characteristics of world-class athletes like a Michael Jordan or Tiger Woods. The same tenacity, drive, desire, and dedication must be applied in business to be successful. There are thousands of ordinary athletes on hundreds of ordinary teams. When individual world-class athletes operate as part of a larger world-class team, they can be hugely rewarded because their fans (customers) cannot go anywhere else to get a better product. The same principle applies to each employee in each company operating at the top of their game. Each employee stands a good chance at improving his or her personal financial position by improving the company's performance.

Take the Long-term View

Why have manufacturing companies in Japan been so successful implementing some of the techniques discussed in this book? It is because they want to provide a safe, secure, and satisfying working environment for coming generations. There is some self interest here: protecting the interests of those who someday will be in the position of protecting you is not a bad idea. Since this book is largely about measurement, look at yourself and measure how well you are doing in this regard. Are you preparing for your future? Are you behaving like a short-term or long-term thinker? The application of Six Sigma and other World-Class manufacturing tools cannot be justified if you are thinking only about this month, this quarter, or this year. True success requires taking the long-term approach.

May all your manufacturing experiences be Lean and operating at Six-Sigma performance levels.

Appendix A

ISO 9000 (2000 Version)

The following is a distilled interpretation of the ISO 9000 (2000 version) requirements, excluding narratives and notes.

- Develop and implement a quality management system.
- Identify the required processes that make up the quality system.
- Describe the quality management processes used to manage the system.
- Ensure that the organization utilizes the quality system process.
- Monitor and manage process performance.
- Improve or modify the quality management system where deemed appropriate.
- Monitor and improve process performance.

DOCUMENT QUALITY SYSTEM

Develop Quality System Documents

- Develop documents to implement the quality system.
- Develop documents that reflect what the organization does.

Prepare Quality System Manual

- Document procedures.
- Describe how processes interact.
- Define the scope of the quality system.

Control Quality System Documents

- Approve documents before they are distributed.
- Provide the correct version of documents at points of utilization.

- Review and re-approve documents whenever they are updated.
- Specify the current revision status of documents.
- Monitor documents that come from external sources.
- Prevent the accidental utilization of obsolete documents.
- Preserve the usability of quality documents.

Maintain Quality System Records

- Utilize records to prove that requirements have been met.
- Develop a procedure to control records.
- Ensure that records are retrievable and useable.

MANAGEMENT REQUIREMENTS

Support Quality

- Promote the importance of quality.
- Promote the need to meet customer requirements.
- Promote the need to meet regulatory requirements.
- Promote the need to meet statutory requirements.
- Develop a quality management system.
- Support the development of a quality system.
- Formulate the organization's quality policy.
- Set the organization's quality objectives.
- Provide quality resources.
- Implement the quality management system.
- Provide resources to implement the quality system.
- Encourage personnel to meet quality system requirements.
- Improve the quality management system.
- Perform quality management reviews.
- Provide resources to improve the quality system.

Satisfy Customers

- Identify customer requirements.
- Expect the organization to identify customer requirements.
- Meet the customers' requirements.

- Expect the organization to meet customer requirements.
- Enhance customer satisfaction.
- Expect the organization to enhance customers' satisfaction.

Establish a Quality Policy

- Define the organization's quality policy.
- Ensure the quality policy serves the organization's purpose.
- Ensure the quality policy emphasizes the need to meet requirements.
- Ensure the quality policy facilitates the development of quality objectives.
- Ensure the quality policy makes a commitment to continuous improvement.
- Manage the organization's quality policy.
- Communicate the policy to the entire organization.
- Review the policy to ensure that it is still suitable.

Carry Out Quality Planning

Formulate Quality Objectives

- Ensure objectives are set for functional areas.
- Ensure objectives are set at organizational levels.
- Ensure objectives facilitate product realization.
- Ensure objectives support the quality policy.
- Ensure objectives are measurable.

Plan Quality Management System

- Plan the development of the quality management system.
- Plan the implementation of the quality management system.
- Plan the improvement of the quality management system.
- Plan the modification of the quality management system.

Control Quality System

Define Responsibilities and Authorities

- Clarify responsibilities and authorities.
- Communicate responsibilities and authorities.

Appoint Management Representative

- Oversee quality management system.
- Report on the status of the quality management system.
- Support the improvement of the quality management system.

Support Internal Communications

- Ensure internal communication processes are established.
- Ensure communication occurs throughout the organization.

Perform Management Reviews

Review Quality Management System

- Evaluate the performance of the quality system.
- Evaluate whether the quality system should be improved.

Examine Management Review Inputs

- Examine audit results.
- Examine product conformity data.
- Examine opportunities to improve.
- Examine feedback from customers.
- Examine process performance information.
- Examine corrective and preventive actions.
- Examine changes that might affect the system.
- Examine previous quality management reviews.

Generate Management Review Outputs

- Generate actions to improve the quality system.
- Generate actions to improve the products.
- Generate actions to address resource needs.

RESOURCE REQUIREMENTS

Provide Quality Resources

- Identify quality resource requirements.
- Identify resources needed to support the quality system.
- Identify resources needed to improve customer satisfaction.
- Provide quality system resources.

- Provide resources needed to support the quality system.
- Provide resources needed to improve customer satisfaction.

Provide Quality Personnel

Utilize Competent Personnel

- Ensure that personnel have the right experience, education, training and skill.

Support Competence

- Define acceptable levels of competence.
- Identify training and awareness needs.
- Deliver training and awareness programs.
- Evaluate effectiveness of training and awareness.
- Maintain a record of competence.

Provide Quality Infrastructure for:

- infrastructure needs,
- building needs,
- workspace needs,
- hardware needs,
- software needs,
- utility needs,
- equipment needs, and
- support services.

Provide Quality Environment

- Identify needed work environment.
- Identify factors needed to ensure products meet requirements.
- Manage needed work environment.
- Manage factors needed to ensure products meet requirements.

REALIZATION REQUIREMENTS

Control Realization Planning

- Plan product realization processes.
- Define product quality objectives and requirements.

- Identify product realization needs and requirements.
- Develop product realization processes and documents.
- Develop product realization record keeping systems.
- Develop methods to control quality during product realization.

Control Customer Processes

Identify Customers' Product Requirements

- Identify the requirements customers want the company to meet.
- Identify the requirements dictated by the product's utilization.
- Identify the requirements imposed by external agencies.
- Identify the requirements the organization wishes to meet.

Review Customers' Product Requirements

- Review requirements before the company accepts orders from customers.
- Maintain a record of product requirement reviews.
- Control changes in product requirements.

Communicate with Customers

- Develop a process to control communications with customers.
- Implement customer communications process.

Control Product Development

Plan Design and Development

- Define product design and development stages.
- Clarify design and development responsibilities and authorities.
- Manage interactions between design and development groups.
- Update design and development plans as changes occur.

Define Design and Development Inputs

- Specify product design and development inputs.
- Record product design and development input definitions.
- Review product design and development input definitions.

Generate Design and Development Outputs

- Create product design and development outputs.
- Approve design and development outputs prior to release.
- Utilize design and development outputs to control product quality.

Carry Out Design and Development Reviews

- Perform product design and development reviews.
- Record product design and development reviews.

Perform Design and Development Verifications

- Carry out product design and development verifications.
- Record product design and development verifications.

Conduct Design and Development Validations

- Perform product design and development validations.
- Record product design and development validations.

Manage Design and Development Changes

- Identify changes in product design and development.
- Record changes in product design and development.
- Review changes in product design and development.
- Verify changes in product design and development.
- Validate changes in product design and development.
- Approve changes before they are implemented.

Control Purchasing Function

Control Purchasing Process

- Ensure purchased products meet requirements.
- Ensure suppliers meet requirements.

Document Product Purchases

- Describe the products being purchased.
- Specify the requirements that must be met.

Verify Purchased Products

- Verify purchased products at company's premises.
- Verify purchased products at suppliers' premises (when required).

Control Operational Activities

Control Production and Service Provision for:

- processes,
- information,
- instructions,
- equipment,
- measurements, and
- activities.

Validate Production and Service Provision

- Prove that special processes can produce planned outputs.
- Prove that process personnel and equipment can produce planned results.

Identify and Track Products

- Establish the identity of product (when appropriate).
- Maintain the identity of product (when appropriate).
- Identify the status of product (when appropriate).
- Record the identity of product (when required).

Protect Property Supplied by Customers

- Identify property supplied to the company by customers.
- Verify property supplied to the company by customers.
- Safeguard property supplied to the company by customers.

Preserve Products and Components

- Preserve products and components during internal processing.
- Preserve products and components during final delivery.

Control Monitoring Devices

- Identify monitoring and measuring needs.
- Identify required monitoring and measuring.
- Select monitoring and measuring devices.
- Select devices that meet monitoring and measuring needs.
- Calibrate monitoring and measuring devices.
- Perform calibrations.
- Record calibrations.
- Protect monitoring and measuring devices from unauthorized adjustment, damage, or deterioration.
- Validate monitoring and measuring software.
- Validate monitoring and measuring software before utilizing it.
- Revalidate monitoring and measuring software when necessary.
- Utilize monitoring and measuring devices.
- Incorporate devices to ensure that products meet requirements.

REMEDIAL REQUIREMENTS

Perform Remedial Processes

- Plan remedial processes.
- Plan how remedial processes will be utilized to assure conformity.
- Plan how remedial processes will be utilized to improve the system.
- Implement remedial processes.
- Utilize remedial processes to demonstrate conformance.
- Utilize remedial processes to improve quality management system.

Monitor and Measure Quality

Monitor and Measure Customer Satisfaction

- Identify ways to monitor and measure customer satisfaction.
- Monitor and measure customer satisfaction.
- Utilize customer satisfaction information.

Plan and Perform Regular Internal Audits

- Set up an internal audit program.
- Develop an internal audit procedure.
- Plan internal audit projects.
- Perform regular internal audits.
- Solve problems discovered during audits.
- Verify that problems have been solved.

Monitor and Measure Quality Processes

- Utilize suitable methods to monitor and measure processes.
- Take action when processes fail to achieve planned results.

Monitor and Measure Product Characteristics

- Verify that product characteristics are being met.
- Keep a record of product monitoring and measuring activities.

Control Nonconforming Products

- Develop a procedure to control nonconforming products.
- Define how nonconforming products should be identified.
- Define how nonconforming products should be handled.
- Identify and control nonconforming products.
- Eliminate or correct product nonconformities.
- Prevent the delivery or utilization of nonconforming products.
- Avoid the inappropriate utilization of nonconforming products.
- Reverify nonconforming products that were corrected.
- Prove that corrected products now meet requirements.
- Control nonconforming products after delivery or utilization.
- Control events when the company delivers or utilizes nonconforming products.
- Maintain records of nonconforming products.
- Describe product nonconformities.
- Describe the actions taken to deal with nonconformities.

Analyze Quality Information

- Define quality management information needs.
- Define the information the company needs to evaluate the quality system.

- Define the information the company needs to improve the quality system.
- Collect quality management system data.
- Monitor and measure the suitability of the quality system.
- Monitor and measure the effectiveness of the quality system.
- Provide quality management information.
- Provide information about customers.
- Provide information about suppliers.
- Provide information about products.
- Provide information about processes.

Make Quality Improvements

Improve Quality Management System

- Utilize audits to generate improvements.
- Utilize quality data to generate improvements.
- Utilize quality policy to generate improvements.
- Utilize quality objectives to generate improvements.
- Utilize management reviews to generate improvements.
- Utilize corrective actions to generate improvements.
- Utilize preventive actions to generate improvements.

Correct Actual Nonconformities

- Review nonconformities.
- Determine causes of nonconformities.
- Evaluate whether the company needs to take corrective action.
- Develop corrective actions to prevent recurrence.
- Take corrective actions when they are necessary.
- Record the results of corrective actions.
- Examine the effectiveness of corrective actions.

Prevent Potential Nonconformities

- Detect potential nonconformities.
- Identify the causes of potential nonconformities.
- Study the effects of potential nonconformities.
- Evaluate whether preventive action is required.
- Develop preventive actions to eliminate causes.

- Take preventive actions when deemed appropriate.
- Record the results that preventive actions achieve.
- Evaluate the effectiveness of preventive actions.

Appendix B

Developing a Concurrent Engineering and Manufacturing Process

The following outline represents the steps taken to develop a concurrent engineering and manufacturing process in a small company. The product is taken from concept launch to preproduction through final design and marketing development. Each company would obviously have to develop their own approach, but this example shows the level of detail needed.

PHASE I—EVALUATION

Preparation of Market Brief

Scope—Establishment of Product Need

Marketing—primary.

- Activity begins with conscious realization that a new product is required based upon verified customer need.
- Evaluation of market potential is based on the strategic plan, competitive strengths and weaknesses, and a data supported estimate of product potential, including any current offerings similar to market share.
- Establish estimated selling price and expected schedule for project margin and sales volume for first five years, including market strategy.
- Complete a "brief" detailing physical and performance requirements and limitations. Define customer needs and expectations and include an analysis of how well competitive products meet these criteria.
- Review brief (new product committee) for input and approval.

Select Team Members—
Assign Project Team Coordinators and Allocate Resources

Scope—Overview of Product Needs and Assignment of Team Members

Having determined that a product need exists, the product manager meets with the steering committee to assign coordinators to the project.

Management—agreement.

- Each departmental manager must review, understand, agree, and sign the market brief.
- The manager assigns a coordinator to the team for their area of responsibility.

Introduce Project to Team—Project Orientation

Engineering manager/marketing manager—primary.

Develop Tentative Schedule for Work Steps for Phase I and Total Project Phases

Scope—Develop a Rough Estimate of Project Time Frame

Project team—primary.

- Establish target dates for major milestones (activities 10, 23, 37, 49, and 55 on project activity chart). (If dates are not provided by marketing brief, then establish.)
- Coordinate plan with engineering/manufacturing/marketing members for phase I.

Develop Design Concepts—
What will the Product Look Like? How will it Work?

Scope—Identify Concepts that will Satisfy Marketing Criteria

Engineer—primary.

- Develop one or more conceptual ideas, which are likely to satisfy the design criteria.
- Concepts are developed using 3D solid modeling for visualization of various perspectives.

- Potential manufacturing and production methods are reviewed with the manufacturing coordinator.
- Prepare report of expected performance capabilities, advantages, and limitations (may include model).

How will the Product be Made? (Manufacturing Methods, Purchased Parts, and Tooling)

Manufacturing/Engineering Coordinator/Purchase Coordinator Determines Methods/Sources (Suppliers may or may not be Involved)

What will the Product Cost?

Scope—Estimate Total Cost for the Product Based on Product Concepts

Project team—primary.

- Evaluate anticipated product costs related to packaging, tooling, labor materials, and testing.

Is the Company Ready? (Evaluate Design Results)

Project team—primary.

- Assemble reported data and prepare proposal in compliance with marketing brief for acceptance.

Present It—Concept Review and Acceptance

Scope—Review Project Need, Product Requirements, Assumptions, and Anticipated Project Costs

Project team—primary.

- Attendees of the meeting are: the general manager, engineering manager, marketing manager, product manager, quality manager, all project team leaders, and anyone else determined to be needed to provide pertinent contributions.

Management—agreement.

- Review product requirements, concepts presented, anticipated sales, project milestone dates, product's proposed configuration, and anticipated project costs.

- If project is acceptable to management for continuation, all signatures should be added to the approval section of the proposal.

Concept/Project Approval

PHASE II—DESIGN

Preliminary CAD Layout

Scope—Prepare Layout Drawings of Major Assemblies

Product engineer—primary.

- The product engineer will prepare adequate layout drawings to analyze performance and verify fit up of component parts.
- All drawings will be produced using CAD in accordance with established drafting standards.
- Layout drawing numbers will be obtained from the product support person in charge of part and drawing numbers.
- Care will be exercised to use existing components wherever possible to minimize inventory requirements.
- Frequent consultations with the assigned manufacturing coordinator or Industrial Engineer will take place to optimize the development phase.
- Frequent consultations will take place with selected suppliers to optimize the development phase.

Preliminary Detail Drawings

Scope—Prepare Required Detail Drawings of Components

Product engineer—primary.

- Using the preliminary layouts as a guide, develop detail drawings of all noncommercial components that are part of the new product.
- All drawings must conform to established drafting standards, which includes critical dimensioning.

Product engineer—agreement.

- When the detail drawings are complete, the product engineer will review them for completeness and accuracy.

Quality review—blueprint review by quality engineering.
Supplier review—concurrent design review by selected suppliers.

Request Quotations—Product

Scope—Obtain Quotes for All New Components of the Product and Supplier Tooling if Appropriate

Product engineer—primary.

- Prepare request for quote form to be sent to the purchasing department along with four sets of the appropriate drawings. Note—the request for quote should identify the anticipated means of producing the components. If multiple means exist, prepare separate requests.
- If there is any chance that the components could be made at another division, a request should be forwarded to industrial engineering to provide quotes.
- For all new purchased components, sample pieces will be ordered to approve the tooling and or vendor processes/quality. Quantity is to be determined and supplied on order.

Purchasing manager—supportive.

- Send drawings to vendors who will provide quotes for the quantities provided on the request for quote form.
- Return a copy of the request for quote form to the requestor indicating the suppliers contacted and the date the prints were sent out.

Capital Equipment/Tooling/Fixture Quotations

Scope—Establish Costs for Equipment Needed In-house to Produce the Proposed Product

Product engineer—supportive.

- Provide the industrial engineer with a complete set of layout drawings, detail drawings, and description of the product/component function.

Industrial engineer—primary.

- Review the layout and component drawings. Communications with the Product Engineer are necessary to establish how the components, subassemblies and final assemblies are to be produced. After evaluating possible alternatives, a method should be established and quotes for the necessary equipment obtained.

- If some manufacturing operations are to be contracted, the industrial engineer is responsible to obtain quotes for these activities. The product engineer must be made aware of this so that he or she can include the added costs as part of the product.

Review and Analyze Quotes for Capital Equipment/Tooling/Fixtures

Scope—Analyze Quote to Determine the Most Cost-effective Way to Produce the Product

Product and industrial engineer—primary.
Product and industrial engineer—supportive.

- Jointly review all quotations received. Consult with purchasing for guidance on likely ordering quantities and desirable suppliers.

Purchasing manager/financial representative/quality representative—supportive.

- Provide information to the product and industrial engineers regarding quality and desirability of various potential suppliers, based on previous experience.

Develop Preliminary Costs

Scope—Determine the Anticipated Manufacturing Cost of the New Product

Product/financial representative—primary.
Industrial engineer/purchasing—supportive.

- Prepare a compilation of expected costs for the product, considering both purchase component costs and the in-house operations costs and/or contract operation costs. This cost should reflect the forecasted volumes established by marketing.

- Review costs with project team. If costs are beyond budget constraints, develop alternative designs/processes to reduce them.

Develop Firm Project Schedule

Scope—Establish a Schedule for Project Completion

Project team—primary.

- Set up a firm schedule for completion of the project using all information determined regarding lead times for materials and tooling. This schedule should also consider in-house workloads and resource availability.

Marketing—agreement.
Sales representative—supportive.

- The marketing coordinator or product manager must participate in scheduling activities to provide information regarding the non-manufacturing functions related to sales literature.
- Sales plan to be developed by sales representative.

Prepare Project-Appropriate Request (PAR)

Scope—Complete Documentation for Financial Approval

Financial representative—primary.
Project team—supportive.

- Using the marketing forecast and cost information provided by the product and industrial engineers, complete the project appropriation request forms.

Build Prototypes/Models

Scope—Fabricate a Model of the Proposed Product for Evaluation

Product engineer—primary.

- Provide guidance in the fabrication of the prototypes to the technician assigned.
- Prototypes need to perform similar to and look like the final product to as great a degree as possible. This may require

contracting with external resources. The product engineer should consult with the industrial engineer and engineering manager if necessary.
- Consult with marketing regarding the important details pertaining to function and aesthetics.

Mechanical Testing of Prototypes

Scope—Perform Preliminary Tests to Determine if the Proposed Product will Meet Design Requirements

Product support engineer—primary.

- Prepare prototypes and test to establish expected capabilities of the product. Any questions relating to the set-up or test procedure should be directed to the product engineer.
- Upon completion of the test, a lab test report must be completed describing the material tested, method of the test, and results.

Product engineer—supportive.

- Provide guidance to the lab technicians regarding test procedures and set-up.
- Review results of test to insure they have achieved the necessary results.
- Prepare a conclusion for the test to be included with the formal test report prepared by the test technician.

Preliminary Testing of Prototypes

Scope—Determine Performance of the Proposed Product

Product support engineer—primary.

- Test the prototype for performance and generate a report to describe the results.

Product engineer—supportive.

- Provide support/guidance to the technician performing the test of the prototype.
- Review the test results to insure the design criteria has been achieved or whether added testing will be required.

- Prepare a conclusion for the lab report indicating whether or not the prototype achieved the intended results and include any recommendations for improvements.

Marketing—agreement.

- The marketing representative should review the results of the test to insure the product will achieve the required performance as specified in the marketing brief.

Prepare Tentative Manufacturing Procedures

Scope—Establish the Probable Method for Producing the Product

Industrial engineer—primary.
Just-In-Time (JIT) cell team members—supportive.

- Using the design layouts, detail drawings, and in consultation with the product engineer, establish a proposed manufacturing procedure, detailing all components to be manufactured and operations to be performed in-house, along with the methods and equipment to be used.

JIT cell team—supportive.

- Works with the industrial engineer to establish the manufacturing procedures, insuring any compromises required do not adversely affect product design criteria.
- If compromise of criteria is necessary, evaluate how the design might be altered to minimize the compromise.

Design Review for Acceptance—Internal Review for Project-Appropriate Request (PAR) Approval

Scope—Design Review to Determine if Project should be Completed or Terminated

Engineering/marketing/manufacturing coordinators—primary.

- After establishing that all activities of this phase have been completed, a meeting is called for review of the project.

Project development team—primary.

- Describes the results of the design program, including important design decisions, compromises, expected product and tooling costs, and schedule for project completion.

Management—agreement.

- Review proposal to insure all processes and activities to this point have been addressed and the costs and proposed methods to produce the new product are acceptable. The introduction date should be considered to determine if all remaining activities can be completed within the time frame allocated. It should be verified that all project goals have been satisfied and that appropriate compromises have been made.

PHASE III—OBTAIN PAR APPROVALS

Finance—primary.
Team—supportive.

- PAR forms are filled out as required with supportive documentation and provided to the controller, general manager, and corporate to secure all necessary approvals to purchase necessary equipment or materials for the project.

Finalize Product Layouts and Detail Drawings

Design—primary.
Purchasing, quality, manufacturing engineering, and production—supportive.

- Layout and detail drawings are provided to team members for approval including labels and customer installation/operation instructions.

Request Catalog Numbers/Determine Marketing Offerings

Marketing—primary.
Team—supportive.

- Determine the catalog numbers to be offered and catalog number logic.

Prepare Preliminary Bills of Material and Structure

Production support engineer—primary.
Manufacturing engineer—supportive.

Prepare Labor Cost Estimates

Manufacturing engineer—primary.
Finance and production control—supportive.

Specify Packaging Needs

Manufacturing engineering—primary.
Supplier and design engineer/marketing—supportive.

- Determine preliminary package requirement.

Review Bills of Material and Release

Design engineer—primary.
Team—supportive.

- Verify completeness and accuracy of bill of material and approve to release.

Detail Marketing/Sales Forecast and Strategy

Design engineer—primary.
Team—supportive.

- Determine warehouse strategy and forecast, as well as sample requirements.

Finalize Quotes

Purchasing—primary.
Team—supportive.

- Following procedure for supplier selection, suppliers for new items are determined and quotes finalized according to quotation procedure.

Open Registrar File and Prepare Submittal

Design engineer—primary.
Team—supportive.

Order Capital Equipment

Manufacturing engineer/purchasing—primary.
Team and supplier(s)—supportive.

Order Component Tooling and Samples

Purchasing—primary.
Design engineer and supplier(s)—supportive.

Approve Tooling Payments

Purchasing—primary.
Finance—supportive.

Tooling Qualification

Quality—primary.
Design engineer/purchasing/supplier(s)—supportive.

- Receive samples, perform measurements to specification, and report deviations to design engineer for review.
- Report acceptance/rejection to purchasing to notify supplier.

Approve Tooling Payments

Purchasing—primary.
Finance—supportive.

Functional Assembly and Design Review

Design—primary.
Team and assemblers—supportive.

Finalize Processing Requirements

Manufacturing engineering—primary.
Team and assemblers—supportive.

- Determine fixturing, training, and other manufacturing needs to support process.

Release Preliminary Product Change Notice (PCN) According to Procedure

Design engineering—primary.
PCN committee—supportive.

Prepare Pilot Run Release—
Prepare First Production Order for Factory/Supplier

Project engineer—primary.
Team—supportive.

- Manufacture—verify completeness of documentation needed for order.

Order Pilot Run Material

Purchasing—primary.
Manufacturing engineering and production control—supportive.

- Place order for pilot run quantity with suppliers. Verify revision level. Notify of anticipated receipt dates and changes thereto.

Finalize Instruction Sheets

Marketing—primary.
Design, quality, and purchasing—supportive.

- Prepare proofs of customer instructions and pictorials to facilitate customer assembly, safe and proper use, including all disclaimers and notices needed. Route for approvals.

Prepare Application Layouts

Marketing—primary.
Sales, design, and quality—supportive.

- Prepare performance charts and other installation data for catalog, cut sheet, and other sales and application engineering support documentation. Review for completeness and accuracy. Route for approvals.

Prepare Catalog Entry and Advertisement

Marketing—primary.
Sales, design, and quality—supportive.

- Prepare proofs of graphics and data for catalog and cut sheet or other sales needs as required for printing. Route for review and approvals.

Safety Test

Engineering—primary.
Team—supportive.

- Perform and report safety test results on samples. Verify limitations are not exceeded.

Other Tests

Engineering—primary.
Team—supportive.

- Perform and report test results on samples.

Shipping/Packaging Test

Manufacturing engineering—primary.
Purchasing, quality, design—supportive.

- Perform and report shipping and packaging tests on samples.

Obtain Registrar Approval

Engineering—primary.
Team—supportive.

- Submit samples and test data with application to registrar. Coordinate with registrar engineer. Notify upon acceptance.

Prepare Manufacturing Procedures

Production engineering—primary.
Team—supportive.

- Prepare written as well as pictorial instructions for proper manufacture (may include posted one-point lectures).
- Verify adequacy and safety of materials and equipment to be used.

Pilot Run

Production—primary.
Team—supportive.

- Produce the first run under actual production conditions. Inspect, observe, and report results. Finalize inspection and test procedures.

Pilot Run Review

Team—primary.
Assemblers and suppliers—supportive.

- Review data and reports of team members from pilot run. Prepare recommendations for any improvements.
- Route for final approval to release for manufacturing and sales order processing.
- Determine and report new product availability details. Verify and report successful compliance with original customer requirements.

PHASE IV—FINAL RELEASE OF NEW PRODUCT

Finalize Revisions to Documentation

Design engineer—primary.
Team and work cell—supportive.

- Review blueprints, reference specifications, posted work instructions, etc., as well as bills of materials. Verify that all changes have been included.

Prepare Release to Production

Product engineer—primary.
Manufacturing/purchasing—supportive.

- Prepare final release document package and distribute using the PCN procedure.
- Management approval is provided by means of signatures at PCN approval. Supplier documentation is updated by purchasing.

Product Launch

Marketing director—primary.
Sales manager—supportive.

- Prepare internal and external pricing guides for the new products including options and accessories according to marketing guidelines. Approve proofs.
- Prepare printed guides for publication and appropriate distribution.

Distribute Literature to Representatives

Marketing director—primary.
Sales manager—supportive.

- Provide list of representatives and arrange to supply them with advertising literature.

Supply Samples to Representatives

Marketing director—primary.
Sales manager—supportive.

- Determine need, cost, and quantity of samples to aid in product sales.
- Determine list of representatives to receive samples.
- Obtain approval for sample order.
- Place sample order(s).
- Build and ship samples.

Production Run

Manufacturing—primary.
Quality/engineering—supportive.

Sales/Customer Service Training

Product Engineer—primary.
Team—supportive.

- Prepare training support materials.
- Schedule presentation with sales/customer service departments. Inform of marketplace considerations, features, construction, use, installation, and benefits of the new product. Provide sales literature.
- Answer technical questions.

Product Introduction

Post-release Review (ISO Required)

Quality—primary.
Marketing—supportive.

- Review product for comparison against original requirements provided in marketing brief.
- Record minutes of the meeting for management review.

Three-Month Audit

Financial manager/quality—primary.
Team—supportive.

- Review sales activity and make recommendations.
- Record minutes for management review.

Six-Month Audit

12-Month Audit

Product Forecast Review and PAR Audit

Financial manager—primary.
Team—supportive.

Appendix C

Example Training Curriculum

The following represents the result of breaking down a skill into trainable and manageable portions by interviewing an expert. This process took about 45 minutes. The purpose of developing a curriculum is to assist the expert in consistently delivering the same information. In the absence of an expert, more detail would be needed including photos, samples, video, etc.

By breaking down the training into modules, the learner can experience the satisfaction of moving from one level to another. Compensation can be tied to demonstrated competency and knowledge of the material.

TOPIC: IN-FEED LAMINATION MACHINE

Skill Level: One

Safety

- Emergency stop buttons;
- Pinch points;
- In-feed rolls;
- Transfer belts;
- Use hoist to eliminate bending as much as possible;
- Band cutting;
- Forklift hazards; and
- Avoid picking up too many pieces at once.

Terminology (How to Identify)

- Work order quantity,
- Reserve quantity,
- Fleece type,

- Glue type,
- Substrate, and
- Veneer.

Tools

- Box knife,
- Moisture detectors, and
- Band cutters.

Paperwork

- Schedule;
- Match part number to substrate and veneer to work order being run;
- Tally rejects on schedule;
- Keep tally of reruns;
- Pull tickets off each load; and
- Complete work order and tally on computer.

Machine Identification

- Hoist;
- U-rack (notify forklift driver of needs);
- In-feed gate system;
- Electric eye (for timing substrate for rolled veneer);
- Jump roll; and
- Emergency stop cable.

Machine Operation

- Need to keep the machine supplied with wood and veneer when running strips;
- Identify the grade zone for each product;
- Watch the counters to ensure correct run quantities;
- Avoid veneer hanging up at jump rolls;
- Assist in cutbacks during set-up;
- Reruns (clicker counter) must be deducted from quantity produced;

- Identify direction in which to feed substrate into machine; and
- Responsibilities when machine is down (breakdown or set-up mode).

Quality Assurance

Grading substrate for:

- Snipe;
- Drags, nicks, chip-out;
- Dents;
- Crook and bow;
- Pitch pockets; and
- Suspect finger joints.

Skill Level: Two

Safety

- How to safely start and stop the machine;
- Handling glue; and
- Material safety data sheets (MSDS).

Tools

- Tape measure,
- Calipers, and
- 17 mm combination wrench.

Paperwork

- Document overages or shortages on ticket;
- Find correct specification sheet from information on schedule (all specifications kept at operator's station);
- List work order number and number of pieces run for each one;
- Part number; and
- Veneer size.

Machine Identification

Emergency stops for:

- In-feed rolls;
- Trim saws; and
- Flying saw.

Machine Operation

- Recognize timing error (electric eyes front and back);
- Perform cutbacks during set-up;
- Interdependency of each operator in the cell;
- Notify operator or cell technician of shortages and overages; and
- Responsibilities when machine is not running.

Quality Assurance

- Check veneer width and thickness against specification; and
- Document measurements on operator's quality assurance log.

Appendix D

101 Things a Six-Sigma Black Belt should Know

*by Thomas Pyzdek**

1. In general, a Six-Sigma black belt should be quantitatively oriented.
2. With minimal guidance, the Six-Sigma black belt should be able to use data to convert broad generalizations into actionable goals.
3. The Six-Sigma black belt should be able to make the business case for attempting to accomplish the goals.
4. The Six-Sigma black belt should be able to develop detailed plans for achieving the goals.
5. The Six-Sigma black belt should be able to measure progress toward the goals in terms meaningful to customers and leaders.
6. The Six-Sigma black belt should know how to establish control systems for maintaining the gains achieved through Six Sigma.
7. The Six-Sigma black belt should understand and be able to communicate the rationale for continuous improvement, even after initial goals have been accomplished.
8. The Six-Sigma black belt should be familiar with research that quantifies the benefits firms have obtained from Six Sigma.
9. The Six-Sigma black belt should know or be able to find the parts per million (PPM) rates associated with different sigma levels (for example, Six Sigma = 3.4 PPM).
10. The Six-Sigma black belt should know the approximate relative cost of poor quality (COPQ) associated with various

sigma levels (for example, three sigma firms report 25% COPQ).

11. The Six-Sigma black belt should know how to quantitatively analyze data from employee and customer surveys. This includes evaluating survey reliability and validity as well as the differences between surveys.
12. The Six-Sigma black belt should understand the roles of the various people involved in change (senior leader, champion, mentor, change agent, technical leader, team leader, facilitator).
13. The Six-Sigma black belt should be able to design, test, and analyze customer surveys.
14. Given two or more sets of survey data, the Six-Sigma black belt should be able to determine if there are statistically significant differences between them.
15. The Six-Sigma black belt should be able to quantify the value of customer retention.
16. Given a partly completed Quality Function Deployment (QFD) matrix, the Six-Sigma black belt should be able to complete it.
17. The Six-Sigma black belt should be able to compute the value of money held or invested over time, including present value and future value of a fixed sum.
18. The Six-Sigma black belt should be able to compute present and future values for various compounding periods.
19. The Six-Sigma black belt should be able to compute the break-even point for a project.
20. The Six-Sigma black belt should be able to compute the net present value of cash flow streams, and to use the results to choose among competing projects.
21. The Six-Sigma black belt should be able to compute the internal rate of return for cash flow streams and to use the results to choose among competing projects.
22. The Six-Sigma black belt should know the COPQ rationale for Six Sigma, that is, he or she should be able to explain what to do if COPQ analysis indicates that the optimum for a given process is less than Six Sigma.

23. The Six-Sigma black belt should know the basic COPQ categories and be able to allocate a list of costs to the correct category.
24. Given a table of COPQ data over time, the Six-Sigma black belt should be able to perform a statistical analysis of the trend.
25. Given a table of COPQ data over time, the Six-Sigma black belt should be able to perform a statistical analysis of the distribution of costs among the various categories.
26. Given a list of tasks for a project, their times to complete, and their precedence relationships, the Six-Sigma black belt should be able to compute the time to completion for the project, the earliest completion times, the latest completion times, and the slack times. He or she also should be able to identify which tasks are on the critical path.
27. Given cost and time data for project tasks, the Six-Sigma black belt should be able to compute the cost of normal and crash schedules and the minimum total cost schedule.
28. The Six-Sigma black belt should be familiar with the basic principles of benchmarking.
29. The Six-Sigma black belt should be familiar with the limitations of benchmarking.
30. Given an organization chart and a listing of team members, process owners, and sponsors, the Six-Sigma black belt should be able to identify projects with a low probability of success.
31. The Six-Sigma black belt should be able to identify measurement scales of various metrics (nominal, ordinal, etc.).
32. Given a metric on a particular scale, the Six-Sigma black belt should be able to determine if a particular statistical method should be used for analysis.
33. Given a properly collected set of data, the Six-Sigma black belt should be able to perform a complete measurement system analysis, including the calculation of bias, repeatability, reproducibility, stability, discrimination (resolution), and linearity.

34. Given the measurement system metrics, the Six-Sigma black belt should know whether or not a given measurement system should be used on a given part or process.
35. The Six-Sigma black belt should know the difference between computing sigma from a data set whose production sequence is known and from a data set whose production sequence is not known.
36. Given the results of an Automotive Industry Action Group (AIAG) Gage Repeatability and Reproducibility study, the Six-Sigma black belt should be able to answer a variety of questions about the measurement system.
37. Given a narrative description of "as-is" and "should-be" processes, the Six-Sigma black belt should be able to prepare process maps.
38. Given a table of raw data, the Six-Sigma black belt should be able to prepare a frequency tally sheet of the data, and to use the tally sheet data to construct a histogram.
39. The Six-Sigma black belt should be able to compute the mean and standard deviation from a grouped frequency distribution.
40. Given a list of problems, the Six-Sigma black belt should be able to construct a Pareto diagram of the problem frequencies.
41. Given a list that describes problems by department, the Six-Sigma black belt should be able to construct a Cross-tabulation and use the information to perform a Chi-Square analysis.
42. Given a table of X and Y data pairs, the Six-Sigma black belt should be able to determine if the relationship is linear or nonlinear.
43. The Six-Sigma black belt should know how to use nonlinearities to make products or processes more robust.
44. The Six-Sigma black belt should be able to construct and interpret a run chart when given a table of data in time-ordered sequence. This includes calculating run length, number of runs, and quantitative trend evaluation.
45. When told the data are from an exponential or Erlang distribution, the Six-Sigma black belt should know that the run chart is preferred over the standard X control chart.
46. Given a set of raw data, the Six-Sigma black belt should be able to identify and compute two statistical measures each for central tendency, dispersion, and shape.

47. Given a set of raw data, the Six-Sigma black belt should be able to construct a histogram.
48. Given a stem and leaf plot, the Six-Sigma black belt should be able to reproduce a sample of numbers to the accuracy allowed by the plot.
49. Given a box plot with numbers on the key box points, the Six-Sigma black belt should be able to identify the 25th and 75th percentile and the median.
50. The Six-Sigma black belt should know when to apply enumerative statistical methods, and when not to.
51. The Six-Sigma black belt should know when to apply analytic statistical methods, and when not to.
52. The Six-Sigma black belt should demonstrate a grasp of basic probability concepts, such as the probability of mutually exclusive events, of dependent and independent events, of events that can occur simultaneously, etc.
53. The Six-Sigma black belt should know factorials, permutations, and combinations, and how to use these in commonly used probability distributions.
54. The Six-Sigma black belt should be able to compute expected values for continuous and discrete random variables.
55. The Six-Sigma black belt should be able to compute univariate statistics for samples.
56. The Six-Sigma black belt should be able to compute confidence intervals for various statistics.
57. The Six-Sigma black belt should be able to read values from a cumulative frequency ogive.
58. The Six-Sigma black belt should be familiar with the commonly used probability distributions, including: hypergeometric, binomial, Poisson, normal, exponential, Chi-Square, Student's t, and F.
59. Given a set of data, the Six-Sigma black belt should be able to correctly identify which distribution should be used to perform a given analysis, and to use the distribution to perform the analysis.
60. The Six-Sigma black belt should know that different techniques are required for analysis depending on whether a given measure (for example, the mean) is assumed known or estimated from a sample. He or she should choose and

properly use the correct technique when provided with data and sufficient information about the data.

61. Given a set of subgrouped data, the Six-Sigma black belt should be able to select and prepare the correct control charts and to determine if a given process is in a state of statistical control.
62. The above should be demonstrated for data representing all of the most common control charts.
63. The Six-Sigma black belt should understand the assumptions that underlie the analysis of variance (ANOVA), and be able to select and apply a transformation to the data.
64. The Six-Sigma black belt should be able to identify which cause on a list of possible causes will most likely explain a non-random pattern in the regression residuals.
65. If shown control chart patterns, the Six-Sigma black belt should be able to match the control chart with the correct situation (for example, an outlier pattern versus a gradual trend matched to a tool breaking versus a machine gradually warming up).
66. The Six-Sigma black belt should understand the mechanics of pre-control.
67. The Six-Sigma black belt should be able to correctly apply exponentially weighted moving average (EWMA) charts to a process with serial correlation in the data.
68. Given a stable set of subgrouped data, the Six-Sigma black belt should be able to perform a complete process capability analysis. This includes computing and interpreting capability indices, estimating the percentage of failures, control limit calculations, etc.
69. The Six-Sigma black belt should demonstrate an awareness of the assumptions that underlie the use of capability indices.
70. Given the results of a replicated 2^2 full-factorial experiment, the Six-Sigma black belt should be able to complete the entire ANOVA table.
71. The Six-Sigma black belt should understand the basic principles of planning a statistically designed experiment. This can be demonstrated by critiquing various experimental plans with or without shortcomings.

72. Given a "clean" experimental plan, the Six-Sigma black belt should be able to find the correct number of replicates to obtain a desired power.
73. The Six-Sigma black belt should know the difference between the various types of experimental models (fixed-effects, random-effects, mixed).
74. The Six-Sigma black belt should understand the concepts of randomization and blocking.
75. Given a set of data, the Six-Sigma black belt should be able to perform a Latin Square analysis and interpret the results.
76. Given a set of data, the Six-Sigma black belt should be able to perform a one way ANOVA, two way ANOVA (with and without replicates), full and fractional factorials, and response surface designs.
77. Given an appropriate experimental result, the Six-Sigma black belt should be able to compute the direction of steepest ascent.
78. Given a set of variables, each at two levels, the Six-Sigma black belt can determine the correct experimental layout for a screening experiment using a saturated design.
79. Given data for such an experiment, the Six-Sigma black belt can identify which main effects are significant and state the effects of these factors.
80. Given two or more sets of responses to categorical items (for example, customer survey responses categorized as poor, fair, good, and excellent), the Six-Sigma black belt will be able to perform a Chi-Square test to determine if the samples are significantly different.
81. The Six-Sigma black belt will understand the idea of confounding and be able to identify which two factor interactions are confounded with the significant main effects.
82. The Six-Sigma black belt will be able to state the direction of steepest ascent from experimental data.
83. The Six-Sigma black belt will understand fold-over designs and be able to identify the fold-over design that will clear a given alias.
84. The Six-Sigma black belt will know how to augment a factorial design to create a composite or central composite design.

85. The Six-Sigma black belt will be able to evaluate the diagnostics for an experiment.
86. The Six-Sigma black belt will be able to identify the need for a transformation in *Y* and to apply the correct transformation.
87. Given a response surface equation in quadratic form, the Six-Sigma black belt will be able to compute the stationary point.
88. Given data (not graphics), the Six-Sigma black belt will be able to determine if the stationary point is a maximum, minimum, or saddle point.
89. The Six-Sigma black belt will be able to use a quadratic loss function to compute the cost of a given process.
90. The Six-Sigma black belt will be able to conduct simple and multiple linear regression.
91. The Six-Sigma black belt will be able to identify patterns in residuals from an improper regression model and apply the correct remedy.
92. The Six-Sigma black belt will understand the difference between regression and correlation studies.
93. The Six-Sigma black belt will be able to perform Chi-Square analysis of contingency tables.
94. The Six-Sigma black belt will be able to compute basic reliability statistics (mean time between failures, availability, etc.).
95. Given the failure rates for given subsystems, the Six-Sigma black belt will be able to use reliability apportionment to set mean-time-between-failure goals.
96. The Six-Sigma black belt will be able to compute the reliability of series, parallel, and series-parallel system configurations.
97. The Six-Sigma black belt will demonstrate the ability to read a Failure Mode and Effects Analysis (FMEA) analysis.
98. The Six-Sigma black belt will demonstrate the ability to read a fault tree.
99. Given distributions of strength and stress, the Six-Sigma black belt will be able to compute the probability of failure.
100. The Six-Sigma black belt will be able to apply statistical tolerancing to set tolerances for simple assemblies. He or

she will know how to compare statistical tolerances to so-called "worst case" tolerancing.

101. The Six-Sigma black belt will be aware of the limits of the Six Sigma approach.

Glossary

14 points: W. Edward Deming's 14 management practices to help companies increase their quality and productivity: 1) create constancy of purpose for improving products and services; 2) adopt the new philosophy; 3) cease dependence on inspection to achieve quality; 4) end the practice of awarding business on price alone; instead, minimize total cost by working with a single supplier; 5) constantly improve every process for planning, production, and service; 6) institute on-the-job training; 7) adopt and institute leadership; 8) drive out fear; 9) break down barriers between staff areas; 10) eliminate slogans, exhortations, and targets for the workforce; 11) eliminate numerical quotas for the workforce and numerical goals for management; 12) remove barriers that rob people of pride of workmanship and eliminate the annual rating or merit system; 13) institute a vigorous program of education and self-improvement for everyone; and 14) put everyone in the company to work to accomplish the transformation.

5-S: A systematic process of workplace organization: sort, set in order, shine, standardize, and sustain.

80/20: Term referring to the *Pareto principle*, first defined by J. M. Juran in 1950, which suggests that 80% of effects come from 20% of possible causes.

A

acceptable quality level (AQL): An ongoing measure of the quality level of a series of lots that, for the purposes of sampling inspection, is the limit of a satisfactory process average.

acceptance sampling: Inspection of a sample from a lot to decide whether to accept that lot. There are two types: *attribute* (the presence or absence of a characteristic), and *variable* (numerical magnitude of a characteristic in comparison to a reference or scale).

acceptance sampling plan: A specific plan that indicates the number of pieces per lot to be sampled and the associated acceptance or non-acceptance criteria to be used.

accreditation: Certification by a duly recognized body, such as the Registrar Accreditation Board. Such bodies accredit those organizations in the business of registering companies that comply to the ISO 9000 series of standards.

activity: Any process, function, or task that occurs over time and has recognizable results. Activities combine to form business processes.

activity-based costing (ABC): An accounting technique that allows an enterprise to determine the actual costs associated with each product and service it produces without regard to the organizational structure of the enterprise or other distorting economic factors.

activity model: A graphic representation that exhibits the activities and interdependencies that make up a business process. An activity model reveals interactions between activities in terms of inputs and outputs while showing the controls and types of resources assigned to each.

American Customer Satisfaction Index (ACSI): Released for the first time in October 1994, this economic indicator is a cross-industry measure of the satisfaction of U.S. household customers with the quality of available goods and services.

analysis of means (ANOM): A statistical procedure for troubleshooting industrial processes and analyzing the results of experimental designs with factors held at fixed levels.

analysis of variance (ANOVA): A statistical technique used to test a hypothesis on the parameters of the model or to estimate variance components. The three ANOVA models are: fixed, random, and mixed.

ANSI: American National Standards Institute

AQP: Association for Quality and Participation

architecture: The organizational structure of a system, identified by its components, their interfaces, and their related methods of execution.

as-is model: A model representing the current condition of an organization without including any specific improvements.
ASME: American Society of Mechanical Engineers
ASQ: American Society for Quality
ASTD: American Society for Training and Development
ASTM: American Society for Testing and Materials
attribute data: Go/no-go information. The control charts based on attribute data can include percent, number of affected units, count, count-per-unit, quality score, and demerit.
autonomation: Stopping a line automatically when a defective part is detected. See *Jidoka*.
availability: The ability of a product or equipment to perform its designated function under stated conditions at a given time. Availability can be expressed by the ratio: uptime ÷ (uptime + downtime), with downtime being when the product or equipment is inoperative (under repair, awaiting spare parts, etc.).
average chart: A control chart in which the subgroup average, X-bar, is used to evaluate the stability of the process level.
average outgoing quality (AOQ): Expected average quality level of outgoing product for a known quality value of incoming product.
average outgoing quality limit (AOQL): Maximum average outgoing quality over all possible levels of incoming quality for a given acceptance-sampling plan and disposal specification.

B

balanced plant: A plant where the capacity of all resources is balanced exactly with market demand.
baseline: The current condition that exists in a situation. Used to differentiate between a current and a future representation.
benchmarking: An improvement process in which a company measures its own performance against other companies viewed as "best-in-class." Undertaken to identify strategies and key characteristics, it uses direct contact, surveys, interviews, technical journals, and marketing materials.
best practice: Identifying a method of accomplishing a business function or process considered superior to all others.

big Q, little Q: Term used to contrast the difference between managing quality in all business processes and products (big Q) and managing quality in a limited capacity—for example, only manufacturing processes (little Q).

blemish: Any noticeable imperfection in a product, but one that should not cause any impairment of its intended normal or reasonably foreseeable use. See also *defect*, *imperfection*, and *nonconformity*.

block diagram: A diagram showing the operation, interrelationships, and interdependencies of components in a system, for example, reliability block diagrams emphasizing only those aspects influencing reliability.

bottleneck: Any resource whose capacity is equal to, or less than the demand placed upon it. See also *constraint*.

brainstorming: A technique that teams use to generate ideas on a particular subject. Each person in the team is asked to think creatively and write down as many ideas as possible for later discussion and review.

BSI: British Standards Institute

business case: A structured proposal for a process-improvement initiative. A business case includes analysis of business process needs or opportunities, proposed solutions, assumptions and constraints, alternatives, costs, proposed benefits, and risks.

business objectives: Goals of the organization that can be measured in some quantitative way.

business-process improvement: The enhancement of an organization's practices through activities used to reduce or eliminate non-value-added activities and costs, while maintaining or improving quality, productivity, and timeliness as evidenced by performance measures.

business-process portal: Portals that bring the right information to the right people at the right time to help them do their work.

business-process redesign (BPR): The transformation of a business process to achieve significant levels of improvement in one or more performance measures relating to fitness for purpose, quality, cycle times, and cost. Uses the techniques of streamlining and removing non-value-added activities and costs.

business-process re-engineering (BPR): A structured effort by all or part of an enterprise to improve the value of products and services while reducing resource requirements. The transformation of a business process seeks to achieve significant levels of improvement in one or more performance measures relating to fitness for use, quality, cycle time, or cost.

C

c chart: Count chart.

calibration: The comparison of a measurement instrument or system of unverified accuracy to a measurement instrument or system of a known accuracy to detect any variation from the required performance specification.

capacity-constraint resources: A series of non-bottleneck resources that act as a constraint, based on the sequence in which work is performed.

cause-and-effect diagram: A graphical display of a detailed list of causes related to a problem or condition for the purpose of discovering its root cause(s) and not just symptoms. Also referred to as the *Ishikawa diagram* (for its developer Kaoru Ishikawa) or *fishbone diagram* because of its resemblance to a fish skeleton.

Chaku-Chaku: A method of conducting single-piece flow, where the operator proceeds from machine to machine, taking the part from one machine and loading it into the next.

change agent: The catalytic force moving firms and value streams out of the world of traditional batch-and-queue and toward a program focusing on a high value-added ratio and exceptional quality.

checklist: A tool used to ensure that all important steps or actions in an operation have been taken. Often confused with check sheet.

check sheet: A simple data-recording device designed to allow the user to readily interpret results. Often confused with checklist.

CMI: Certified mechanical inspector

common causes: Causes of variation inherent in a process over time.

company culture: A system of values, beliefs, and behaviors inherent in a company that is established or reinforced by top management.

conformance: A positive indication or judgment that a product or service has met the requirements of a specification, contract, or regulation.

constraint: Anything that limits a system from achieving higher performance or throughput, or a process that severely limits the organization's ability to achieve desired performance.

continuous process improvement: The ongoing improvement of products, services, or processes through incremental and breakthrough improvements. A policy that encourages, mandates, and/or empowers employees to find ways to improve process and product performance on a continual basis.

control chart: A chart with upper and lower control limits plotting values for a key statistical measure or subgroup and used to detect trends toward either control limit.

corrective action: The implementation of solutions to reduce or eliminate an identified problem.

cost of poor quality (COPQ): The costs associated with providing poor-quality products or services. These include: *internal failure costs* (costs recognized before the customer receives the product); *external failure costs* (costs of defects found after the customer receives the product); *appraisal costs* (costs to sort or determine the degree of nonconformance); and *prevention costs* (costs incurred to eliminate failures).

cost of quality (COQ): Actually the cost of *poor* quality, a term coined by quality guru Philip Crosby.

count chart: A control chart for evaluating the stability of a process.

covariance: The impact of one variable upon others in the same group.

CQA: Certified quality auditor

CQE: Certified quality engineer

CQI: Continuous quality improvement

CQM: Certified quality manager

CQT: Certified quality technician
CRE: Certified reliability engineer
CSQE: Certified software quality engineer
cumulative-sum control chart: A control chart showing the cumulative sum of deviations of successive samples from a target value. Each plotted point represents the algebraic sum of the previous plotted point and the most recent deviations from the target.
customer satisfaction: Delivering a product or service that meets customer requirements. Customer delight is achieving the goal of delivering a product or service that exceeds customer expectations.
customer/supplier partnership: A long-term relationship between a buyer and supplier characterized by teamwork, collaboration, and mutual confidence.

D

decision matrix: A matrix used by teams to evaluate problems or alternative solutions. Teams rate possible solutions on a scale of one to five for each criterion and the ratings of all the criteria for each possible solution are added to determine its total score. These scores are then used to help decide which solution deserves the most attention. As an alternative, a weighted value can be assigned to each option to avoid assigning too much importance to any one criteria.
defect: Any indication of a product's non-fulfillment of an intended requirement or reasonable expectation for use. *Class 1, very serious*, can lead to severe injury or catastrophic economic loss; *Class 2, serious*, can lead to significant injury or significant economic loss; *Class 3, major*, relates to major problems with respect to an intended use; and *Class 4, minor*, relates to minor problems with respect to intended use.
demerit chart: A control chart for evaluating a process in terms of a demerit or quality score.
Deming cycle: See *plan-do-check-act cycle*.
Deming Prize: Award given annually to organizations that have successfully applied companywide quality-control initiatives and demonstrated the capability to sustain these improvements.

Deming, W. Edwards: Consultant, teacher, and author on the subject of quality. After sharing his expertise in statistical process quality control to help the U.S. during World War II, Deming taught the Japanese SPC methodologies to help that nation recover. In a career spanning five decades, Deming brought his 14-point program to many major corporations.

dependability: The degree to which a process, machine, or product is operable and capable of performing its required function at any randomly chosen time. Dependability can be expressed by the ratio: time available ÷ (time available + time required).

dependent events: Events that can only occur after a previous event or prerequisite.

Design of Experiments (DoE): A subset of statistics dealing with planning, conducting, analyzing, and interpreting controlled tests to evaluate factors or inputs that control the value or outputs of a parameter, group of parameters, or process.

diagnostic journey and remedial journey: A two-phase investigation used by teams to solve chronic quality problems. In the diagnostic journey, a team journeys from the symptom of a chronic problem to its cause. In the remedial journey, a team journeys from the cause to its remedy.

discounted cash flow: A method of performing an economic analysis that takes the time value of money into account. Used to remove interest rates and inflation factors from a calculation so that the results from various analyses can be more accurately compared.

E

economic analysis: A systematic method of comparing and selecting between alternative methods of accomplishing a set objective; testing the assumption, constraints, cost, and benefits of each alternative to find which will most likely result in the optimum result.

employee involvement: A practice within an organization whereby employees regularly participate in making decisions on how their work areas operate, including making suggestions for improvement, planning, goal setting, and monitoring performance.

empowerment: A condition whereby employees have the authority to make decisions and take action in their work areas without prior approval. For example, any operator can stop the production process if he or she perceives a problem.

enfranchisement: A condition where the lines of authority to make decisions are clearly defined. A manager does not impose decisions on employees where they should have the responsibility. Team members do not unreasonably expect their every whim and wish to be granted. Ideas from teams are provided in the form of recommendations. Mutual respect and two-way communication are practiced daily.

enterprise: An organization that exists to perform a specific mission and achieve associated goals and objectives.

experimental design: A formal plan that details the specifics for conducting an experiment; that is, which parameters, factors, and tools are to be used.

external customer: A person or organization who receives a product, service, or information, but is not part of the organization supplying it.

evaporating clouds: A term for the problem-solving method used in the Theory of Constraints that is essentially the same as conflict resolution.

F

failure-mode analysis (FMA): A method of determining which malfunction symptom appears immediately before or after the failure of a system or failure of a critical parameter in that system. After examining each potential cause for each symptom, the product is redesigned to eliminate the problems.

Failure-Mode and Effects Analysis (FMEA): A procedure in which each potential failure mode in every sub-item of an item is analyzed to determine its effect on other sub-items and on the required function of the complete item.

fishbone diagram: See *cause-and-effect diagram*.

fitness for use: A term used to indicate that a product or service meets the customer's needs and defined purpose.

fixed cost: Costs that do not vary with the level of production and remain even if an activity or process stops.

flow chart: A graphical representation of process steps used for better understanding.

force-field analysis: A technique for analyzing the forces that aid or hinder an organization in reaching an objective. An arrow pointing to an objective is drawn down the middle of a piece of paper. The factors that will aid the objective's achievement, called the *driving forces*, are listed on the left side of the arrow. The factors that will hinder its achievement, called the *restraining forces*, are listed on the right side of the arrow.

funnel experiment: An experiment that demonstrates the effects of tampering. Marbles are dropped through a funnel in an attempt to hit a flat-surfaced target. It is used to show that adjusting a stable process to compensate for an undesirable result can produce output that is worse than if the process had been left alone. Alternative: dropping playing cards onto a target.

G

gage repeatability and reproducibility (GR&R): Evaluation of an instrument's accuracy by determining whether its measurements are repeatable and reproducible.

Gantt chart: A time-phased project plan. A type of bar chart used in process planning and control to display planned work, interrelated work, and finished work along a time line.

geometric dimensioning and tolerancing (GD&T): A method to minimize production costs by showing dimensions and tolerancing on a drawing while considering part-feature variations and relationships.

go/no-go: A gaging or inspection method wherein only two parameters are possible: *go* (conforms to specifications), or *no-go* (does not conform to specifications).

H

Heijunka: Maintaining manufacturing volume as constant as possible. Same as production smoothing.

histogram: A graphic representation of variation in a set of data.

Hoshin Kanri: Closely related to policy deployment. It involves selection of goals (and projects to achieve those goals), designation of people and resources for project completion, and establishment of project metrics.

Hoshin planning: Breakthrough planning. A strategic planning process in which a company develops multiple vision statements indicating where the company should be in the next five years. Goals and work plans are then developed based on the vision statements. Routine audits are used to monitor progress.

I

imperfection: Departure from the intended level of conformance.

in-control process: A process in which the statistical measure being evaluated is in a state of statistical control.

inspection: Measuring, examining, testing, or gaging one or more characteristics of a product, and comparing the results with specified requirements to determine a level of conformity.

internal customer: The person, process, or department receiving output (product, service, or information) from another person, process, or department within the same organization.

internal setup: Die-setup procedures that must be performed while a machine is stopped.

Ishikawa diagram: See *cause-and-effect diagram.*

Ishikawa, Kaoru: Pioneer in quality control activities in Japan. Developed the *cause-and-effect diagram* in 1943.

ISO: International Organization for Standardization

ISO 9000: Quality management and assurance standards adopted by ISO (International Organization for Standardization, founded 1947), which are based on an international consensus of over 110 countries. First published in 1987, it has since been adopted as a national standard in more than 80 countries.

J

Jidoka: The management philosophy to stop a production line automatically when a defective part is detected.

judgment inspection: A form of inspection used to determine nonconforming product.

Juran, Joseph M.: Founder of the Juran Institute and noted author on quality-management technologies.

JUSE: Union of Japanese Scientists and Engineers

Just-In-Time (JIT): A method of requesting the delivery of material, products, or services at the moment they are needed in an effort to reduce inventory, queue time, storage space, handling, and degradation.

K

Kaikaku: Radical improvement, usually applied only once within a value stream.

Kaizen: Based on the Japanese root *kai*, meaning alter or change, and *zen*, good, it has become the theme for all continuous-improvement programs.

Kanban: Basically a signal to make something or move something in response to a downstream "pull" demand. The signal can be as simple as a transfer slip or empty bin.

Kano model: A visual representation of the relationship between quality and a customer's level of satisfaction.

knowledge management: Leveraging collective wisdom to increase responsiveness and innovation.

L

leadership: Essential to any improvement effort and accomplished through establishing a vision, communicating that vision, and providing tools, resources, and knowledge necessary to accomplish the vision.

Lean Manufacturing: Producing sellable products or services at the lowest operational cost, while optimizing inventory levels. Put simply: getting more done with less.

load-load: A method of conducting single-piece flow, where the operator proceeds from machine to machine, taking the part from one machine and loading it into the next.

lot: Defined quantity of units per sales order, work order, sample, or Kanban.

lower control limit (LCL): Control limit for points below the central line in a control chart.

M

maintainability: The probability that a maintenance action can be performed within a stated time when performed under stated conditions using stated procedures. *Serviceability* is the ease of conducting scheduled inspections and servicing. *Repairability* is the ease of restoring service after a failure.

Malcolm Baldrige National Quality Award (MBNQA): Award established by Congress in 1987 to raise awareness of quality management and to recognize U.S. companies that have implemented successful quality-management systems. Named after the late Secretary of Commerce.

management systems: Software tools for supporting the modeling, analysis, and enactment of business processes.

mean time between failures (MTBF): The average time interval between failures, measured in hours, cycles, miles, etc.

MIL-Q-9858A: Military standard for quality program requirements.

MIL-STD: Military standard

MIL-STD-105E: Military standard for sampling procedures (inspection by attributes).

MIL-STD-45662A: Military standards for creating and maintaining a calibration system for measurement and test equipment.

muda: Japanese term for any human activity that absorbs resources, but creates no value.

multivariate control chart: Control chart used in evaluating process stability of two or more variables or characteristics.

N

n: Sample size. The number of units in a sample.

Nagara system: Production system where unrelated tasks can be performed by the same operator at the same time or within the Takt time cycle.

NIST: National Institute of Standards and Technology

nominal group technique (NGT): Structured brainstorming technique allowing a team to quickly come to consensus on the importance of issues, problems, or solutions.

nonconformity: Nonfulfillment of a specified requirement.

nondestructive testing and evaluation (NDT): Testing and evaluation methods that do not damage or destroy the product being tested (for example, x-ray, fluorescent penetrant, etc.).

non-value-added (NVA): Activities or actions taken that do not add value to the product or service and may not have a valid business reason for being performed. There are various types of NVA activities including those required to meet governmental regulations. Same as muda.

O

one-touch exchange of dies (OTED): First promoted by Japan's Shigeo Shingo at Toyota, refers to the reduction of die set-up to a single step. Off-line set-up may still play a part in preparing for a changeover where machine downtime is minimized through the efforts of an off-line set-up team.

operating expenses: The money required for a system to convert raw material or inventory into sellable product.

organization diagnostics: Process of identifying organizational problems with processes, procedures, technology, culture, etc.

out-of-control process: Process in which the statistical measure being evaluated is not in a state of statistical control.

P

Pareto chart: Graphical tool for ranking causes from most significant to least significant, which is named after 19th-century economist Vilfredo Pareto who suggested that most effects come from relatively few causes (80% of the effects come from 20% of the possible causes).

percent (p) chart: Control chart for evaluating the stability of a process.

performance measure: Indicator used to evaluate quality, cost, or cycle-time characteristics of an activity or process against a target value or standard.

physical transformation task: Taking a specific product from raw materials to finished product to the hands of the customer.

pitch: Pace and flow of a product through production.

plan-do-check-act (PDCA) cycle: Four-step process for quality improvement. 1) A plan to effect improvement is developed. 2) The plan is carried out, preferably on a small scale. 3) The effects of the plan are observed. 4) The results are studied to determine what was learned and what can be predicted. Sometimes referred to as the Shewhart cycle or Deming cycle.

Poka-Yoke: Mistake proofing. Designing a process so the chance of an operator or machine error is significantly reduced or completely eliminated.

policy deployment: Selection of goals and projects to achieve those goals. Designation of people and resources to ensure project completion. Establishment of project metrics.

prevention versus detection: Term used to contrast two quality approaches. Prevention refers to the approach designed to prevent nonconformances. Detection refers to the approach designed to detect nonconformances already in products or services.

process-capability index: A value assigned to a process describing its demonstrated ability to maintain a tolerance specified for a characteristic.

process Kaizen: Continuous improvement through incremental improvements.

process model: Also activity model. Graphic representation of a business process that exhibits the activities and their interdependencies. The model reveals interactions between activities in terms of inputs and outputs while showing the controls and types of resources assigned to each one.

product or service liability: The obligation of a company to make restitution for loss related to personal injury, property damage, or other harm caused by its product or service.

production smoothing: Keeping total manufacturing volume (flow) as constant as possible. Same as *Heijunka*.

Q

quality: A subjective term for which each person or customer may have their own definition. Characteristics of a product or service that impact its ability to satisfy stated or implied needs. A product or service that is free of nonconformities.

quality assurance (QA)/quality control (QC): Terms meant to imply confidence, a state of certainty, consistency, or to describe a managed system designed to minimize unexpected variation in product or service quality.

quality audit: Systematic examination and review (generally performed by an independent party) to determine whether quality activities and related results comply with planned arrangements.

quality circles: Also called *quality-control circles*. Quality improvement groups composed of a small number of employees—10 or fewer—along with a supervisor or facilitator meeting on a routine basis to focus on improvement opportunities. Originated in Japan.

quality engineering: The process of analyzing all stages of a manufacturing system to maximize the quality and outputs of the process.

quality-function deployment (QFD): Using a cross-functional team to reach consensus that final engineering specifications will meet the needs of the customer, the QFD process is often referred to as listening to the "Voice of the Customer." QFD is a structured method in which customer requirements are translated into appropriate technical requirements for each stage of product development and production.

quality score chart (Q chart): Control chart for evaluating the stability of a process in terms of a quality score. The quality score is a weighted sum of the count of events of various classifications with each classification assigned a weight.

quality trilogy: A three-pronged approach to managing quality: quality planning, quality control, and quality improvement.

quick changeover: The ability to change tooling and fixtures rapidly, so multiple products can be run in quick succession on the same machine.

R

random sampling: Common sampling technique in which sample units are selected in such a manner that all units under consideration have an equal chance of being selected.

range chart (R chart): Control chart in which a subgroup range (the difference between the largest and smallest value), R, is used to evaluate the variability within a process.

real value: Attributes and features of a product or service for which a customer is willing to pay.

Registrar Accreditation Board (RAB): A board evaluating the competency and reliability of registrars—independent third party auditing firms that assess and register companies as complying to the appropriate ISO 9000 series standards.

regression analysis: Statistical technique for determining the most descriptive mathematical expression for the functional relationship between one response and one or more independent variables.

reliability: The probability of a product performing its intended function (without failure) for a stated period of time and under stated conditions.

right-sized equipment: Matching tooling and equipment to the job and space requirements of Lean production.

right the first time: Concept that it is always beneficial and more cost-effective to take the necessary steps up front to ensure a product meets its requirements rather than to provide a product that will likely require rework later.

robust: A product or process design that can remain stable (with minimum variation) even though influenced by factors that may constantly change.

run chart: A statistical tool to identify the number of parts or subgroups trending in a particular direction either above, below, or through the process mean.

S

scatter diagram: Graphical technique to analyze the relationship between two variables. Plotted on a simple X and Y axis, it helps visualize the effect of one variable upon another.

seven tools of quality: Basic TQM tools in the world-class toolbox. The tools are the: cause-and-effect diagram, check sheet, control chart, flow chart, histogram, Pareto chart, and scatter diagram.

Shewhart cycle: See *plan-do-check-act cycle.*

Shewhart, Walter A.: Considered the father of statistical quality control, he described the basic principles of this new discipline in his book, *Economic Control of Quality of Manufactured Product* (American Society for Quality Control, 1980).

Shingo, Shigeo: Japanese engineer credited with developing much of the Toyota production system for fast changeovers (SMED) and foolproof systems (Poka-Yoke).

single-minute exchange of dies (SMED): Goal for the reduction in die set-up time. Although set-up in a single minute is not often attained, single-digit exchange of dies (SDED) is a goal within many make-to-order shops. Goals and methods were originally developed by Shigeo Shingo.

Six-Sigma quality: Term used to indicate that a process is controlling defects or errors to parts-per-million levels.

special causes: Causes of variation that arise because of special or unusual circumstances not inherent to the process.

standard-deviation chart (s chart): Control chart where subgroup standard deviation, *s*, is used to evaluate the stability of the variability within a process.

standard work: Specific tasks assigned to get a job done in a specified amount of time while ensuring that the job is done correctly each and every time.

statistical fluctuations: Information that cannot be precisely predicted.

statistical process control (SPC): Application of statistical techniques to control a process.

statistical quality control (SQC): Application of statistical techniques to control quality. SQC includes acceptance sampling as well as statistical process control.

stove pipes: Also known as departmental silos. Term is commonly used to reflect that a business function operates in a vertically integrated manner and does not efficiently or effectively interact with related functions.

structural variation: Variation caused by routine, predictable, systematic changes, such as seasonal patterns and long-term trends.

sub-optimization: Where improvements made in one activity result in a negative impact or are offset by losses in another activity.

subprocesses: Series of combined operations.

supplier quality assurance: Confidence that a supplier's product or service will fulfill its customers' needs. It is based on defined product and quality program requirements, evaluation of alternative suppliers, supplier selection, joint quality planning, cooperation with the supplier during the execution of the contract, proof of conformance to requirements, certification of qualified suppliers, and creation and use of supplier-quality ratings.

T

Taguchi method: Introduced by Dr. Genichi Taguchi, this set of quality-control concepts (based on an audio analogy) encourages a high degree of focus on efforts to identify design and process parameters that will minimize the chance of "loss" due to process "noise" (process variability).

Takt time: Daily production number required to meet customer demand (orders in hand) divided into the number of working hours in the day.

Theory of Constraints (TOC): A management philosophy that stresses removal of constraints to increase velocity and throughput, while decreasing inventory and operating costs.

throughput: The rate at which money (sales) is generated through the system.

to-be model: Models for the future state, developed by brainstorming and applying improvement opportunities to the current state (as-is) business environment.

total quality management (TQM): A management approach focusing on long-term success through customer satisfaction. It is built on the idea of participation from all members of the organization, with everyone focusing on improving processes, products, services, and the culture.

trend control chart: Control chart in which a deviation of any subgroup average (X-bar) from an expected trend is used to evaluate the stability of a process.

Type I error: An incorrect decision to reject something when it is actually acceptable.

Type II error: An incorrect decision to accept something when it is actually unacceptable.

U

u chart: Count-per-unit chart.

upper control limit (UCL): Control limit for points above the central line in a control chart.

V

value-added activity: Adding value to an output product or service. An activity or process that changes the product into something a customer can use and is willing to pay for.

value-adding process: Activities that transform inputs into customer-usable outputs.

value-mapping analysis: Analyzing the value stream to identify the ratio of value-added to non-value-added activities.

variable cost: Cost elements that vary with the amount of product or service produced. Variable costs go to zero if an activity stops.

variables data: Measurement information rather than judgment or attribute information. Control charts based on variables data include: average (X-bar); range, R; and sample standard deviation, s.

variation: A change in data, characteristic, or function caused by either special causes, common causes, tampering, or structural variation.

visual controls: Displaying the status of an activity so workers can see it and take appropriate action.

vital few, useful many: Term that describes the Pareto principle, the 80/20 rule. (The second half of the term originally was referred to as the "trivial many," but realizing that there are no trivial problems, it has been changed to "useful many.")

W

waste: See also *muda*. Anything that consumes resources, but does not add value to the product.

work flow: A business process beginning with a commitment and ending with the termination of the commitment, where systems, elements, or activities are related to one another by a trigger relation or triggered by external events.

work-flow-management systems: Integrated software tools used to support the modeling, analysis, and enactment of business processes.

World Class: A term used to indicate a standard against which all others are measured: best of the best.

X

X-bar chart: Statistical-average chart.

Y

yield: The number of acceptable products produced, related to those expected or scheduled to be produced.

Z

zero defects: The goal that, even though mistakes are made, workers can move closer to the goal of zero defects by watching critical details, avoiding errors, and analyzing and controlling all of a system's subsystems and lower-level products.

Bibliography

Akao, Y. 1991. *Hoshin Kanri: Policy Deployment for Successful TQM*. Portland, OR: Productivity Press.

Allen, J., Robinson, C., and Stewart, David, eds. *Lean Manufacturing: A Plant Floor Guide*. Dearborn, MI: Society of Manufacturing Engineers (SME).

Ammerman, Max. 1998. *The Root Cause Analysis Handbook—A Simplified Approach to Identifying, Correcting, and Reporting Workplace Errors*. Portland, OR: Productivity, Inc.

Arnold, Kenneth L. 1994. *The Managers Guide to ISO 9000*. New York: The Free Press.

ASQ. 1988. *Solving Quality and Productivity Problems: Goodmeasure's Guide to Corrective Action*. Milwaukee, WI: American Society for Quality (ASQ).

Brown, Mark Graham. 2001. *Baldrige Award Winning Quality: How to Interpret the Baldrige Criteria for Performance Excellence*. Milwaukee, WI: American Society for Quality.

Cohen, Lou. 1995. *Quality Function Deployment*. Upper Saddle River, NJ: Prentice Hall.

Conner, Gary. 2001. *Lean Manufacturing for the Small Shop*. Dearborn, MI: Society of Manufacturing Engineers (SME).

Covey, Stephen R. 1989. *Seven Habits of Highly Effective People*. New York: Simon & Schuster.

Day, Ronald G. 1993. *Quality Function Deployment: Linking a Company with its Customers*. Milwaukee, WI: American Society for Quality.

Dettmer, H. William. 1997. *Goldratt's Theory of Constraints: A Systems Approach to Continuous Improvement*. Milwaukee, WI: Quality Press.

Dodson, Bryan, Keller, Paul A., and Pyzdek, Thomas. 2001. *Six Sigma Study Guide*. Tucson, AZ: Quality Publishing.

Eckes, George. 2001. *Making Six Sigma Last: Managing the Balance Between Cultural and Technical Change*. New York: John Wiley & Sons.

Fisher, Dennis. 1995. *The Just-in-Time Self-Test: Success Through Assessment and Implementation*. New York: McGraw-Hill.

Goldratt, Eliyahu M. 1994. *It's Not Luck*. Great Barrington, MA: North River Press.

——. 1999. *Theory of Constraints*. Great Barrington, MA: North River Press.

Groover, Mikell P. 2000. *Automation, Production Systems, and Computer-Integrated Manufacturing*. Upper Saddle River, NJ: Prentice Hall.

Hales, H. Lee, Andersen, Bruce and Fillmore, William E. 2002. *Fundamentals of Manufacturing Cell Planning* (textbook, workbook, and video program). Dearborn, MI: Society of Manufacturing Engineers (SME).

Hirano, Hiroyuki. 1996. *5S for Operators: 5 Pillars of the Visual Workplace*. Portland, OR: Productivity Press.

——. 1989. *JIT Implementation Manual*. Portland, OR: Productivity Press.

Imai, Masaaki. 1986. *Kaizen: The Key to Japan's Competitive Success*. New York: McGraw-Hill.

Jordan, James A., Jr. and Michel, Frederick J. *The Lean Company: Making the Right Choices*. Dearborn, MI: Society of Manufacturing Engineers (SME).

Juran Institute. 2001. *The Six-Sigma Basic Training Kit: Implementing Juran's 6-Step Quality Improvement Process and Six-Sigma Tools*. New York: McGraw-Hill Professional Publishing.

Kaydos, Wilfred J. 1991. *Measuring, Managing, and Maximizing Performance: What Every Manager Needs to Know About Quality and Productivity to Make Real Improvements in Performance*. Cambridge, MA: Productivity Press.

Krener, Chuck, Rizzuto, Ron, and Case, John F. 2000. *Managing by the Numbers: A Commonsense Guide to Understanding and Using Your Company's Financials*. Cambridge, MA: Perseus Press.

Liker, Jeffery K. 1998. *Becoming Lean*. Portland, OR: Productivity Press.

Louis, Raymond. 1997. *Integrating Kanban with MRPII*. Portland, OR: Productivity Press.

Mahoney, Francis X. and Thor, Carl G. 1994. *The TQM Trilogy: Using ISO 9000, the Deming Prize, and the Baldrige Award to Establish a System for Total Quality Management*. New York: Amacom.

McClave, James T. and Dietrich, Frank H., II. 1991. *Statistics*. Indianapolis, IN: Macmillan Publishing Co.

Mickalski, Walter J. 1997. *Tool Navigator*. Portland, OR: Productivity Press.

Monden, Yasuhiro. 1998. *Toyota Production System*. Norcross, GA: Institute of Industrial Engineers.

Muther, Richard. 1979. *Systematic Layout Planning*. Kansas City, MO: Management and Industrial Research Publications (MIRP).

Nakajima, Seiichi. 1988. *Introduction to TPM*. Portland, OR: Productivity Press.

Nyman, L., ed. *Making Manufacturing Cells Work*. Dearborn, MI: Society of Manufacturing Engineers (SME).

Pande, Peter, Neuman, Robert, and Cavanagh, Roland. 2001. *The Six-Sigma Way Team Field Book: An Implementation Guide for Process Improvement Teams*. New York: McGraw-Hill.

Peters, Thomas and Waterman, Robert, Jr. 1982. *In Search of Excellence: Lessons From America's Best-Run Companies*. New York: Harper and Row.

Phillips, E. 1997. *Manufacturing Plant Layout.* Dearborn, MI: Society of Manufacturing Engineers (SME).

Productivity Development Team. 1996. *Mistake-Proofing for Operators: The ZQC System* (Shop Floor Series). Portland, OR: Productivity Press.

Productivity Press Team. 1999. *Cellular Manufacturing: One-Piece Flow for Work Teams* (Shop Floor Series). Portland, OR: Productivity Press.

Pyzdek, Thomas. 2001. *The Six-Sigma Handbook.* New York: McGraw-Hill.

Rath & Strong. 2001. *Six-Sigma Pocket Guide.* Lexington, MA: Rath & Strong.

Regan, Michael D. 2000. *The Kaizen Revolution.* Raleigh, NC: Holden Press.

Robinson, Charles J. and Ginder, Andrew P. 1995. *Introduction to Implementing TPM: The North American Experience.* Portland, OR: Productivity Press.

Rother, M. and Harris, R. 2001. *Creating Continuous Flow: An Action Guide for Managers, Engineers, and Production Associates.* Brookline, MA: Lean Enterprise Institute.

Rother, Mike and Shook, John. 1998. *Learning to See.* Brookline, MA: Lean Enterprise Institute.

Roy, Ranjit K. 1990. *A Primer on the Taguchi Method.* Dearborn, MI: Society of Manufacturing Engineers (SME).

——. 2001. *Design of Experiments Using the Taguchi Approach: 16 Steps to Product and Process Improvement.* New York: Wiley-Interscience.

Scholtes, Peter R. 1998. *The Team Handbook.* Madison, WI: Joiner Associates.

Schonberger, Richard J. 1982. *Japanese Manufacturing Techniques: Nine Hidden Lessons in Simplicity.* New York: Free Press.

Sekine, Kenichi and Keisuke, Arai. 1992. *Kaizen for Quick Changeover: Going Beyond SMED.* Portland, OR: Productivity Press.

Shillito, M. Larry. 2000. *Acquiring, Processing, and Deploying Voice of the Customer*. Boca Raton, FL: CRC Press.

Shingo, Shigeo. 1985. *A Revolution in Manufacturing: The SMED System*. Portland, OR: Productivity Press.

——. 1986. *Zero Quality Control: Source Inspection and the Poka-Yoke System*. Portland, OR: Productivity Press.

Shulyak, Lev. 1997. *40 Principles: TRIZ Keys to Technical Innovation*. Worcester, MA: Technical Innovation Center.

Simchi-Levi, David, Kaminski, Philip, and Simchi-Levi, Edith. 1999. *Designing and Managing the Supply Chain: Concepts, Strategies, and Cases*. New York: Irwin/McGraw-Hill.

SME. 1988. *Implementing Just-in-Time* (*Manufacturing Insights* video). Dearborn, MI: Society of Manufacturing Engineers (SME).

——. 1990. *Layout Improvements for Just-in-Time* (*Manufacturing Insights* video). Dearborn, MI: Society of Manufacturing Engineers (SME).

——. 1990. *Setup Reduction for Just-in-Time* (*Manufacturing Insights* video). Dearborn, MI: Society of Manufacturing Engineers (SME).

——. 1990. *Total Quality Management* (*Manufacturing Insights* video). Dearborn, MI: Society of Manufacturing Engineers (SME).

——. 1991. *Flexible Small Lot Production for JIT* (*Manufacturing Insights* video). Dearborn, MI: Society of Manufacturing Engineers (SME).

——. 1992. *TQM: Creating a Culture of Continuous Improvement* (*Manufacturing Insights* video). Dearborn, MI: Society of Manufacturing Engineers (SME).

——. 1992. *TQM: The First Steps* (*Manufacturing Insights* video). Dearborn, MI: Society of Manufacturing Engineers (SME).

——. 1992. *TPM: Total Productive Maintenance* (*Manufacturing Insights* video). Dearborn, MI: Society of Manufacturing Engineers (SME).

——. 1993. *Managing Teams in Manufacturing* (*Manufacturing Insights* video). Dearborn, MI: Society of Manufacturing Engineers (SME).

——. 1993. *Rapid Response Manufacturing* (*Manufacturing Insights* video). Dearborn, MI: Society of Manufacturing Engineers (SME).

——. 1994. *Benchmarking Manufacturing Processes: Your Practical Guide to Becoming Best in Class* (video program). Dearborn, MI: Society of Manufacturing Engineers (SME).

——. 1994. *Cellular Manufacturing in a Global Marketplace* (video program). Dearborn, MI: Society of Manufacturing Engineers (SME).

——. 1994. *Total Productive Maintenance in America* (video program). Dearborn, MI: Society of Manufacturing Engineers (SME).

——. 1995. *Implementing QS-9000* (video). Dearborn, MI: Society of Manufacturing Engineers (SME).

——. 1996. *Mistake Proofing: Achieving Zero Defects* (video). Dearborn, MI: Society of Manufacturing Engineers (SME).

——. 1997. *Customer-Focused Manufacturing* (video). Dearborn, MI: Society of Manufacturing Engineers (SME).

——. 1997. *Work Measurement* (*Manufacturing Insights* video). Dearborn, MI: Society of Manufacturing Engineers (SME).

——. 1998. *Fundamentals of Plant Floor Layout* (video program). Dearborn, MI: Society of Manufacturing Engineers (SME).

——. 1999. *Continuous Improvement: Sustaining the Effort* (video). Dearborn, MI: Society of Manufacturing Engineers (SME).

——. 1999. *Measurement and Gaging* (video). Dearborn, MI: Society of Manufacturing Engineers (SME).

——. 1999. *Supply Chain Management* (*Manufacturing Insights* video). Dearborn, MI: Society of Manufacturing Engineers (SME).

——. 1999. *Theory of Constraints: Meeting Customer Demand with Synchronized Production* (video). Dearborn, MI: Society of Manufacturing Engineers (SME).

——. 2000. *Introduction to Lean Manufacturing* (*Manufacturing Insights* video). Dearborn, MI: Society of Manufacturing Engineers (SME).

——. 2000. *Lean Manufacturing at Miller SQA* (*Manufacturing Insights* video). Dearborn, MI: Society of Manufacturing Engineers (SME).

——. 2000. *Lean Manufacturing at TAC Manufacturing* (*Manufacturing Insights* video). Dearborn, MI: Society of Manufacturing Engineers (SME).

——. 2000. *Mapping Your Value Stream* (*Manufacturing Insights* video). Dearborn, MI: Society of Manufacturing Engineers (SME).

——. 2000. *Quick Changeover for Lean Manufacturing* (*Manufacturing Insights* video). Dearborn, MI: Society of Manufacturing Engineers (SME).

——. 2000. *Visual Controls* (*Manufacturing Insights* video). Dearborn, MI: Society of Manufacturing Engineers (SME).

——. 2001. *Making Lean Happen in Your Organization*; *Developing a Lean Supply Base*; *Lean Culture*; *Business Culture and the Lean System*; *The Challenge of Staying Lean After Initial Application*; *Six-Sigma Improvement Program Deployment in a High-Speed Process Environment*; *Leading a Six-Sigma Launch*; *Roadmap to Successful Six-Sigma Implementation*; *Staying Out of the Hands of Charlatans*; *Using Quality Methodology Throughout the Commercialization Process*. Videotaped presentations from the SME 2001 Annual Meeting & Manufacturing Leadership Forum. Dearborn, MI: Society of Manufacturing Engineers (SME).

——. 2001. *Six Sigma* (*Manufacturing Insights* video). Dearborn, MI: Society of Manufacturing Engineers (SME).

Smith, Dave. 1991. *Quick Die Change*. Dearborn, MI: Society of Manufacturing Engineers (SME).

Smith, Wayne. 1998. *Time Out: Using Visible Pull Systems to Drive Process Improvement*. New York: John Wiley & Sons.

Stamatis, D.H. 1995. *Failure Mode and Effects Analysis: FMEA from Theory to Execution*. Milwaukee, WI: American Society for Quality.

Suzaki, Kiyoshi. 1987. *The New Manufacturing Challenge*. New York: The Free Press.

——. 1989. *The New Manufacturing Challenge: Techniques for Continuous Improvement* (video program). Dearborn, MI: Society of Manufacturing Engineers (SME).

——. 1992. *The New Shop Floor Management: Empowering People for Continuous Improvement* (video program). Dearborn, MI: Society of Manufacturing Engineers (SME).

Tennant, Geoff. 2001. *Six Sigma: SPC and TQM in Manufacturing and Services*. Brookfield, VT: Gower Publishing Co.

Tincher, Michael G. 1996. *High-Velocity Manufacturing*. Gurnee, IL: Buker.

Torres, Cresencio and Fairbanks, Deborah M. 1996. *Team-Building: The ASTD Trainer's Sourcebook*. New York: McGraw-Hill.

Winchell, W. *Inspection and Measurement in Manufacturing*. Dearborn, MI: Society of Manufacturing Engineers (SME).

Womack, James P., Jones, Daniel, and Roos, Daniel. 1991. *The Machine That Changed the World*. New York: Harper Collins.

Index

D

E

F

G

H

I

J

K

L

M

N

O

P

Q

R

S

T

U

V

W

X

CD-ROM INSTALLATION INSTRUCTIONS

Recommended Requirements

Hardware: Pentium processor, 64 MB RAM, 24 MB of available hard-disk space, 8× CD-ROM drive or faster. *Operating System:* Microsoft Windows® 95, 98, ME, NT 4.0 SP5, 2000, or later. *Software:* Internet Explorer® 5.0 or later, Netscape® 4.7 or later; Microsoft (1997 or later) Excel®, Word®, and Powerpoint®; and Windows® Media Player®.

Running the CD

This CD is designed to automatically play when inserted in your CD-ROM drive. Adobe Acrobat Reader® must run from the CD to view the content of the CD-ROM. If AUTORUN has been disabled on your computer, the CD will not automatically play. In this case, you can boot Acrobat Reader from the CD. The path is X:\Reader\AcroRd32.exe (where X is the drive letter of your CD-ROM drive). Double click on "AcroRd32.exe" to launch Acrobat Reader. In Acrobat Reader, go to <File>Open>X:\SixSigma\main.pdf to open the beginning document.